Christa Reicher, Fabio Bayro-Kaiser,
Päivi Kataikko-Grigoleit, Sarah Müller,
Jan Polívka (Eds.)

Urban Integration

Stadt- und Raumplanung
Urban and Spatial Planning

Band/Volume 19

LIT

URBAN INTEGRATION

From Walled City to Integrated City

edited by

Christa Reicher, Fabio Bayro-Kaiser,
Päivi Kataikko-Grigoleit, Sarah Müller,
Jan Polívka

LIT

This book is printed on acid-free paper.

Bibliographic information published by the Deutsche Nationalbibliothek
The Deutsche Nationalbibliothek lists this publication in the Deutsche Nationalbibliografie; detailed bibliographic data are available in the Internet at http://dnb.dnb.de.

ISBN 978-3-643-91179-7 (pb)
ISBN 978-3-643-96179-2 (PDF)

A catalogue record for this book is available from the British Library.

Flössergasse 10
CH-8001 Zürich
Tel. +41 (0) 76-632 84 35
E-Mail: zuerich@lit-verlag.ch http://www.lit-verlag.ch
Distribution:
In the UK: Global Book Marketing, e-mail: mo@centralbooks.com
In North America: Independent Publishers Group, e-mail: orders@ipgbook.com
In Germany: LIT Verlag Fresnostr. 2, D-48159 Münster
Tel. +49 (0) 2 51-620 32 22, Fax +49 (0) 2 51-922 60 99, e-mail: vertrieb@lit-verlag.de
e-books are available at www.litwebshop.de

Translations and proof reading: Claudia Becker
Graphic recording: Antje Rieder
Layout: Päivi Kataikko-Grigoleit

Many thanks to the universities, students, project partners and supporters involved. Without them, this exciting project would not have been possible.

The project was funded by the Federal Ministry of the Interior, Building and Community within the framework of the National Urban Development Policy, by Deekeling Arndt Advisors in Communications GmbH and by ILS – Research Institute for Urban and Regional Development.

Index

2. Resilience + Diversity & Inclusion

Preface

Worldwide, cities and entire regions are affected by structural change and they face the challenge of comprehensive transformation processes. It is their international interrelation, the global context, and their urban-regional rescaling and interrelation that are new about these processes. Such has also been visible in the Ruhr area since the last decades. With the end of the German coal-industry in December 2018, an era of 150 years of industrial history came to an end; an era associated with extraordinary prosperity and fundamental social achievements. This marks a historical caesura for millions of people in the coal-mining districts. Phenomena such as globalization and digitalization accelerate this change and they bring – once again - several aspects of life into motion. Such developments might trigger a promising dynamic if used in a smart way and with a surplus for local people, their living environment and the regional economy. Moreover, as the urbanization is constantly, though slowly, changing with societal shifts – and with the internationalization and regionalization as the two current major trends that accelerate this pace – the aspect of integration becomes of importance. The crucial task becomes how to integrate the formerly industrially coordinated pace of societal change which meanwhile became a matter of fundamental diversification processes. In a post-industrial, widely neo-liberal society, the social economic coherence and its spatial patterns remain an extremely important, though not self-evident prerequisite for a stable future. This can be promoted or obstructed by broadly defined societal and spatial conditions.

As a third milestone in the Transforming City Regions network, the "Urban Integration: From Walled City to Integrated City" initiative, together with partners from science and practice, will deepen the challenges of integration in the context of design and composition of neighbourhoods, cities and regions. The altered framework conditions, the increasing urbanization pressure, growing social polarization and the current phenomena of digitalization shall be particularly considered. They challenge both citizens and key players in new ways, and are therefore to be examined also in regard to the existing settlement patterns and the planning management.

Questions towards integration are at the centre: do spatial conditions and spatial concepts contribute to the promotion of urban integration? By which strategies for implementation should they be underpinned? To these questions, spatial and strategic answers in urban design and urban development should follow. This shall be done through discussing existing approaches and concepts in the region's framework, and through debating them with international experts and their expertise on the relevance and implementation of integration in their specific urban contexts.

Aims of the symposium are to

(1) underpin the relevance of the topic of "integration" in an international perspective and with that, draw conclusions for the Ruhr region,

(2) expand the knowledge of national and international best-practice-projects to "city and integration",

(3) build a platform for an international exchange of science and practice regarding questions of transformation,

(4) strengthen the pioneer role of the Ruhr region with focus on the project "Glückauf Zukunft", a follow-up project for a post-mining region, in an international perception.

The major topic of transformation as well as the focus on integration are of global relevance. The framework of the symposium not only allows an exchange of expertise, but also pursues a learning process allowing to profit from the knowledge of others. In the global world of today, this might be quite normal in general, but is not yet established as self-evident in particular planning contexts. In doing so, the international perception of the Ruhr region as a forward-looking laboratory for the design of transformations processes should be further strengthened.
The ongoing research on the topic of transforming city regions already has been one of our main issues during the last years. After the workshops and symposiums in New York (USA) and Amman (Jordan), the international symposium "Urban Integration: From Walled City to Integrated City" in May 2018 at the Zollverein World Heritage Site in Essen (Germany) is the fourth milestone along our way, and the second in a row at Zollverein, where, back in 2015, the topic of internationalization in connection with the regional urban design was initially addressed in this format.

Christa Reicher for the "Transforming City Regions" team
Dortmund, June 2019

I. Transforming City Regions: Urban Integration in an International Perspective

Introduction

Worldwide, metropolitan regions are changing rapidly to absorb the urban growth expected over the next 20-30 years. While urban transformations in the global south will be dominated by irregular processes of urbanization, the formal regulatory frameworks operating in the global north must also be interrogated to ensure a more sustainable, just and vibrant urban future. Framed by the paradigm shift of the digital revolution, the anticipated impacts of climate uncertainty, and the aftermath of the post-industrial socio-economic regimes, time calls for innovative strategic responses. The complexity of these challenges requires an integrated and transdisciplinary approach that allows for diverse, inclusive and multi-scalar strategies. With an emphasis on urban integration, this symposium approaches the agglomeration of the Ruhr region as a breeding ground for urban experimentation through the consideration of four principals:

Polycentricity

From a spatial point of view, polycentric structures of urban agglomeration - like the Ruhr area - are a promising model for sustainable urban and regional development. Rather than focusing on individual cities, polycentricity emphasizes the connectivity and synergy between urban centres, and the role of smaller units such as neighbourhoods and urban quarters. It also requires innovative approaches in land management and land use. An integrated urban development, therefore, must develop strategies to strengthen the identity of - and connectivity between - cities, urban quarters and neighbourhoods on different spatial and functional levels.

Multi-scale spatial models

Uneven growth simultaneously occurring at regional and neighbourhood scales requires a new model for integrated design able to negotiate a multiplicity of nested scales. Multi-scalar systems are marked by locally divergent development speeds and perspectives within the same agglomeration. Multi-scale models, therefore, need to integrate diversity in spatial and functional dynamics and responsiveness to address the effects of growth and depopulation, centralization and peripherization, and a range of environmental, socio-political and economic dynamics, and break it down to the topic of settlement space management. A new multi-scale model requires innovative mechanisms to link legal frameworks, facilitate new forms of knowledge creation at every planning level, and promote exchange among all stakeholders. The discussion of best practice case studies will expand the knowledge base and provide further insights for the development of new strategies.

Resilience

The resilience of cities depends upon the flexibility and adaptive capacity of their strategic infrastructure and planning systems and their socio-ecological components. Key conditions are: redundancy, social cohesion, cultural diversity, and governance transparency and responsiveness. Until recently, the guiding principle of sustainable development has been to achieve stable ecological function. However, profound changes to the human, environmental and technical systems, and its functions, are affected now by heterogeneity, disturbance, and conditions of uncertainty - new parameters for the future institutions of civil society.

Diversity and inclusion

The phenomenon of migration is accelerating change and generating new social formations. In an increasingly complex urban realm, how can urban design processes, policies, and regulations guiding development support integration? From the perspective of the individual and the formation of new community structures, by designing inclusive mechanisms and novel institutions to enable the representation and participation of diverse publics. A starting point for new ideas and interventions is a local example - the 2017 „Modellquartier Integration“ project in the Ruhr area.

1. Polycentricity + Multi-Scale Spatial Models

Infra Eco Logi Urbanism

Geoffrey Thün, Kathy Velikov

That ninety percent of the world's population will be living in cities by 2050 is by now a familiar prognosis on future urbanization. While some global cities prepare for unprecedented hyperdensity, in many parts of the world this urban migration is producing the development of continuous low density urban development between city centres; multicentred continuous agglomerations of built form, also known as megaregions. These polycentric conurbations, the spatial products of the ad-hoc and emergent forces of mobile capital, infrastructure, logistics and environmental management, confound traditional questions of urban boundaries, identities, politics, natural and manmade systems. The emergence of megaregions also signals a moment where reconsideration of the material and cultural commons enables new conceptions of resources, publics, distribution and rights. Within these contexts, design, and specifically architecture, acquires new agency.

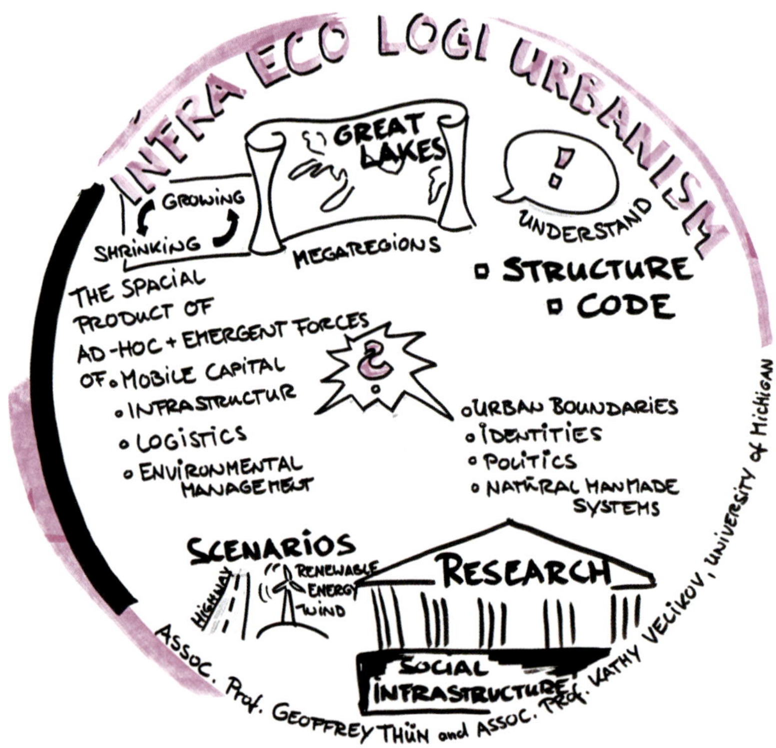

Infra Eco Logi Urbanism is a multi-year design research project undertaken by the experimental practice RVTR that posits a manifesto for architecture at the regional scale. A projected future development for the Great Lakes Megaregion of North America is the situational subject within which speculative design work positions approaches to urban analysis and design intervention that aim to reconceptualise future urban ecologies, cross-border governance, politics, infrastructure and public architecture.

The working methodology of the project moves through multiple scales of consideration in both analysis and design, and combines contemporary perspectives of ecology, geo-design, assemblage and actor-network theory as well as utopian paradigms. The work has been developed through three parallel streams of research. One: an intensive regional analysis undertaken within a methodological framework of system (shed cartographies of interdependent systems), structure (typological physical forms and artefacts) and code (operational practices and rule sets). Two: an assembly of historic disciplinary influences within the topics of fragmentary utopias, urban mega forms and large-scale urban interiors. Three: a speculative design proposal for the megaregion developed through systematic infrastructural interventions and detailed architectural designs which leverage sustainable energy networks toward new infrastructural architectures. This project advocates for the potential for architecture to not only engage in urban processes but also for the potential of urban form to imagine alternative futures by structuring spaces for negotiation and public action at the scale of the megaregion.

RVTR's Infra Eco Logi Urbanism, a major solo travelling exhibition, was exhibited at the Centre de Design de l'UQAM, Montreal (2013), Paul Crocker Gallery, Ryerson University, Toronto (2014), Paul Rudolph Hall Gallery, Yale University New Haven, CT (2014), LRA Gallery, University of Michigan, Ann Arbor, MI (2015), and the Ewing Gallery, University of Tennessee, Knoxville, TN (2015). A monograph of the project, Infra Eco Logi Urbanism: A Project for the Great Lakes Region, was published by Park Books, Zurich in 2015.

Dynamics of Periphery: A Cross-Scale Approach

Jörg Schröder

Dynamics of Periphery offers an unusual reading and a cross-cutting approach towards topics that are discussed in several spatial and sectorial frameworks. Both in Europe and as part of global debates, aims of integrative and balanced development of metro-regional peripheries and of areas outside new metropolitan regions call for a fresh perspective in research, planning practice, and politics. New metropolitan bodies and cooperations in fluid and changing constellations include spatial phenomena that in the past have been characterized as peripheral in different meanings and contexts, oscillating between positive connotations of territorial and cultural diversity and — in many cases monodirectionally coined — economic and social criticalities. The range of phenomena considered here refers to spatial terminologies such as rural areas, small and medium-sized towns, suburbs, tourist and leisure zones, and extended infrastructural areas.

Today, in perspectives of post-metropolitan conditions, fragmentation, segregation, or new polarities, the approach of Dynamics of Periphery envisions a relational understanding of proactive aspects in the non-core and non-central situations that have been neglected in urban research and theory for many

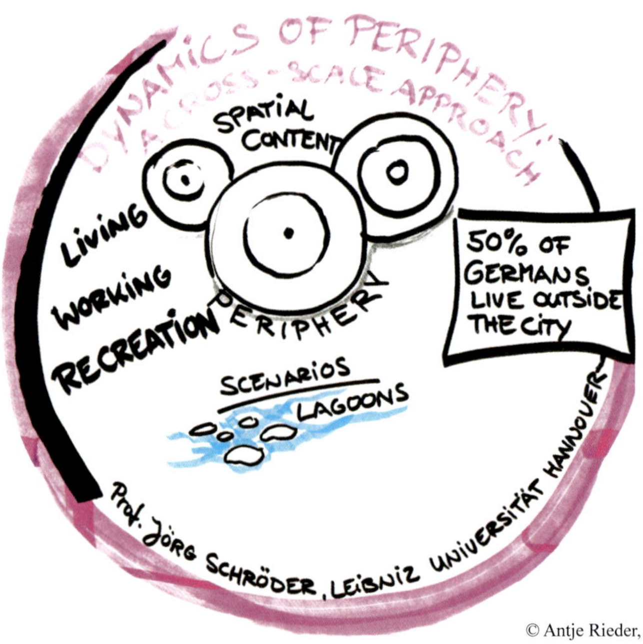

© Antje Rieder, Graphic Recording

years: due to orientations towards abstract hybridity or illusionary "flatness" on the one hand, or due to a focus on exclusive "global hubs" on the other hand. Though, the terms of polarities, networks, and "spikyness" may offer access also to current spatial processes, not only in micro-scales, but for an overall cross-scale perspective in analysis and concept-drawing, by highlighting spatial, social, cultural, ecological, and economic interdependencies.

Dynamics of Periphery — as international platform funded by DAAD — proposes a methodology for this change of perspective: first, from impulses and possible dynamics originating from the margins and borders of metropolitan constellations that lead also to a rethinking of planning and design scenarios for city and territory in general; and second, from the inclusion of a prospective mode through an architectural-urbanistic approach that is linked to re-evaluation of peripheries. Effectively, a focus on design — revisiting the concept of Gestaltung as vivid impulse but also as hubris of modernity — can open up new insights in spatial analysis, for concepts and strategies, and in governance processes. The aim of Dynamics of Periphery is to accelerate this debate through exploring its interdependencies to the paradigms of resilient, inclusive, and productive cities, to successful projects of pro-active public bodies and engaged citizens, and to inventions in urbanistic methodology. The focus on a Cross-Scale Approach — answering to demands of effective experience of positive spatial change as well as to the call for more strategic and long-term efforts in climate change and global competitiveness — is intended to show chances and limitations of a renewed role of creative expertise and design-based innovations in addressing the future of metro-regional dimensions.

Several themes and topographies of Dynamics of Periphery, insights from research projects in Germany, Italy, and in the Alpine Space, are examined towards three transversal topics for possible trans-scalar approaches: In-transition Processes; Place: Managing the Change; Narratives: Storytelling. New interactive, integrative, shared, and responsive modes of Gestaltung in urbanism and territorial development are discussed, leading also to the question how they influence debates on academic and professional abilities and the general mission of urbanistic study and research programmes. A Cross-Scale Approach is advocated for a refreshed awareness of logics of context and of scales, hence of a relational expertise.

Bibliography

Schröder J., Carta, M., Ferretti, M., Lino, B. (Ed.) 2018: Dynamics of Periphery. Jovis Verlag GmbH, Berlin

Funded by DAAD (German Academic Exchange Service) University Dialogue Southern Europe, financed by the German Foreign Office.

Flexible Green Spaces

Stephanie Haury

Initial position

Through re-urbanization processes more and more people live in urban areas. The move to the big cities continues. In Germany, the city centres are currently the big winners, while the borders of the cities tend to stagnate and the peripheral areas have even to accept population losses (BBSR, 2015a: 11). By 2035, the population in growing areas of the federal territory will continue to increase, despite the decreasing total population (BBSR, 2015b: 11). In particular, in the large metropolitan areas, it will, therefore, come to redensification for which green areas have to be abandoned. In regions with population growth and economic dynamism, undeveloped areas are, therefore, under development pressure, as profitability is often preferred. The municipal challenge is to harmonize a high degree of structural density and the widest possible variety of different uses in compact areas.

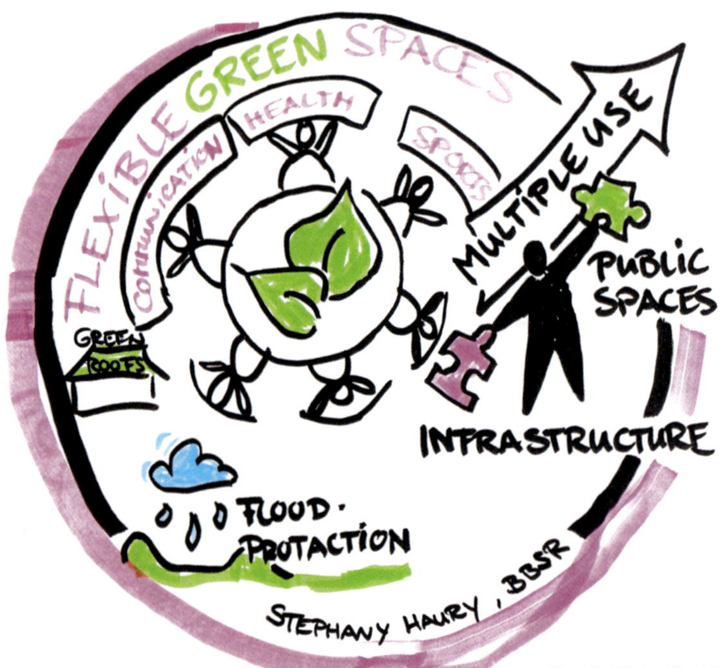

© Antje Rieder, Graphic Recording

Green infrastructure

Green spaces have a special status in cities due to their ability to incorporate many functions. They are classified under the term „urban green infrastructure". This urban green infrastructure delivers important contributions to the area of quality of life, recreation, enhancement of the image, ecological-climatic balance, biodiversity and health. At the same time, it shapes the city and forms part of the Baukultur. Green infrastructure also contributes to improving the resilience of cities, in particular to environmental, climate and health risks.

Damage caused by heavy and extreme rainfall in cities will continue to increase in the future. „Green and open spaces are therefore of great importance as flood plain areas and for the creation of a time delay of drainage. Urban green increases the climate resilience, i.e. the resilience to climate change-related risks, and is significantly less expensive than technical solutions for retaining large quantities of water" (BMUB, Grünbuch Stadtgrün, 2015: 73).

Multiple use

In addition to the general importance of green spaces, multi-use and multi-coding are of great importance for the resilience of cities. This includes a superposition of the functions of green spaces. Green spaces thus serve e.g. at the same time for water management (flood areas) and are temporarily used for sports or recreational areas. As a result, aspects of Baukultur can be strengthened if rainwater retention basins are no longer pure concrete basins, but designed spaces for living.

Multiple uses are playing an increasingly important role not only in growing regions. Even in economically weak regions with shrinking processes, these can be interesting approaches to superimpose different uses in tight budgets because multifunctional approaches always lead to area and cost savings and guarantee the best possible use of space.

Requirements for implementation

In the implementation of multiple uses, the challenge for administrations is to build a collaborative planning culture and to incorporate sectoral planning. Departmental responsibilities and various legal bases have to be brought together and have to be coordinated. In this coordination process, questions about financing, traffic safety, hygiene, safety, planting, furnishing and storage height have to be clarified.

The subject of multiple use is not a perennial favourite in planning law, but rather a one-day fly. For zoning plans legal norms must be clearly defined in terms of content. The definitions and other statements must be clear and unambiguous. By means of this requirement of certainty, additional provisions must be foreseen when setting multiple uses. The planning law allows so-called „dependent stipulations", which regulate the execution of the type of use to superimpose. Thus, different „executions" of green areas such as a „park use" can be combined with "flood protection measures". This should be done by drawing and textual stipulations. New determination types may not arise because section 9 (1) BauGB (German Federal Building Code) contains a final catalogue for determinations. Alternatively, municipalities and private owners can conclude "urban planning contracts". The BBSR has had a legal review of this as part of a research project (BBSR, 2015c: 50). However, this is not an easy task for the municipalities. Especially in those municipalities that are not sufficiently equipped with staff capacities.

More of that!

There is a great variety for freedom of use for open spaces. Well-known examples are the Benthemplein in Rotterdam, the Südliche Gartenstadt in Potsdam-Bornstedt or the areas of the Cloudburst Management Plan in Copenhagen. The goal must be to implement models based on local requirements that create added value for the municipalities and their specialized administrations and for the citizens. It is, therefore, important to strengthen the multi-functional concepts needed in the sense of resilience. Because „there are still many untapped potentials in the cities for multiple uses of space in order to open up new free space qualities despite the increasing density" (BBSR, 2018: 100). For this reason, the Federal Government will examine the effectiveness of the development of „multicoding strategies" for urban green as a measure of the White Paper „Stadtgrün" published in 2017, and will develop a guideline out of this examination.

Bibliography

Bundesinstitut für Bau-, Stadt- und Raumforschung (BBSR) (Ed.) 2015a: Divergenzen und Konvergenzen in Großstadtregionen – kleinräumige Analysen. BBSR-Analysen KOMPAKT 07/2015, Bonn

Bundesinstitut für Bau-, Stadt- und Raumforschung (BBSR) (Ed.) 2015b: Die Raumordnungsprognose 2035 nach dem Zensus. BBSR-Analysen KOMPAKT 05/2015, Bonn

Bundesministerium für Umwelt, Naturschutz, Bau und Reaktorsicherheit (BMUB) (Ed.) 2015: Grün in der Stadt – Für eine lebenswerte Zukunft. Grünbuch Stadtgrün, Berlin

Hessisches Ministerium für Wirtschaft, Verkehr und Landesentwicklung (Ed.) 2012: Freiräume entwickeln - Lebensräume schaffen, Wiesbaden

Bundesinstitut für Bau-, Stadt- und Raumforschung (BBSR) (Ed.) 2015c: Überflutungs- und Hitzevorsorge durch die Stadtentwicklung, Bonn

Bundesinstitut für Bau-, Stadt- und Raumforschung (BBSR) (Ed.) 2018: Urbane Freiräume, Bonn

Multi-Level, Multi-Scale Urban Models

Michael Wegener

Urban development is a multi-level process embedded in global, national and regional development and is itself multi-scale ranging down to the neighbourhood and building level, and so involves a hierarchy of jurisdictions and decision-making spheres: Economic development of cities is co-determined by trends and policy decisions at the global, European and national level, spatial development by policy decisions about taxation, infrastructure investments and planning regulations at the national or regional level. Demographic and social development of cities results from decisions of households and firms at the neighbourhood level, and mobility from decisions on activities and communication by households and individuals. This requires multi-level planning support systems embracing all relevant spatial levels.

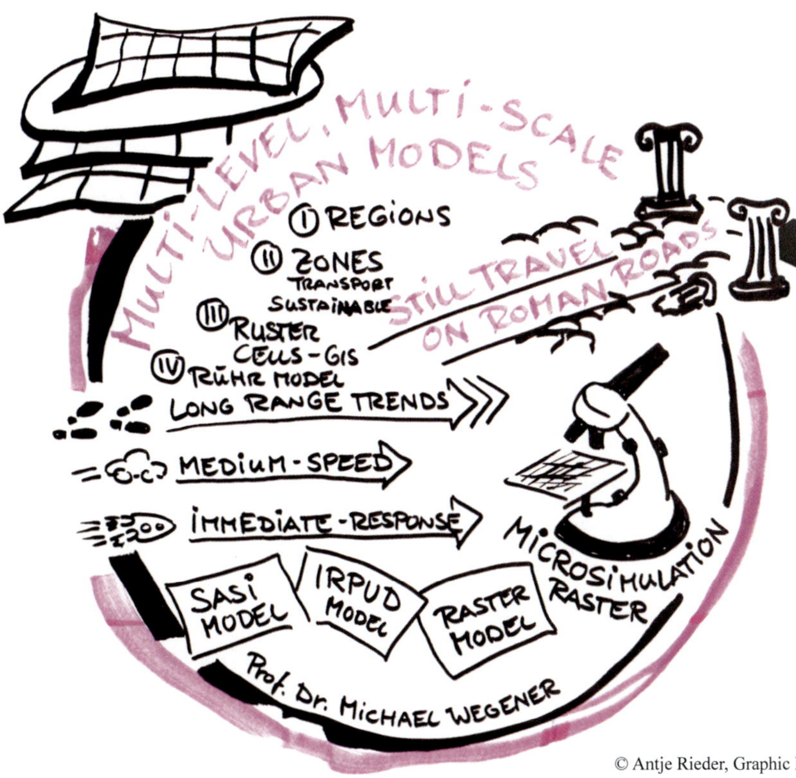

In addition urban models need to address the different speeds of urban change processes: The physical structure of cities changes only slowly over centuries – we still travel on Roman roads. Economic and demographic development occurs over years and mobility within a day. This implies that urban models have to incorporate processes with different speeds ranging from long-range trends with response times of generations via medium-speed dynamics to immediate-response equilibrium. In addition new data sources and computer techniques have stimulated a trend towards ever more disaggregation in space and time, and this, too, requires different resolutions at different spatial levels.

The reasons for multi-level, multi-scale models become even more relevant if ongoing substantial changes in the political context of urban development at the European, national, regional and local scale are considered. To cope with the challenges of new technologies, energy scarcity, climate change and refugee immigration, urban models will need to be able to predict the decision behaviour of firms, households and individuals in cities subject to policy decisions at different levels of spatial and temporal resolution:

(1) Growing interconnectedness of economic development, transport and energy provision will require modelling the long-term impacts of policy decisions, such as cohesion or transport policies, at the European or national level.
(2) Energy scarcity and climate protection policies, such as emission standards, taxation, fuel price regulation or charging infrastructure will require modelling the long-term impacts of decisions at the European, national or regional level.
(3) Climate change adaptation measures, such as flood prevention and management and development control in flood plains and fresh-air corridors will require medium-term policy decisions at the national or regional level.
(4) Increasing pressure of refugee immigration from Africa and Middle East countries caused by poor economic decisions, poverty and climate change will require short-term political action at the regional or local scale.
(5) Behavioural and technological trends already now apparent, such as telework, online shopping, electro mobility, car-sharing, autonomous cars and renewable energy will require new models with different time scales at the regional or local scale.
(6) Growing concern for quality of life, air quality and noise exposure will require microscopic modelling at the raster or building level.

These new developments will have impacts on the priorities and standards of urban modelling. As the changes will be so large, it will be necessary to focus less on detail and more on basic essentials. Moreover, data collected in the past will become of less value for explaining behaviour after fundamental changes,

so extrapolation will become more and more useless. This implies that behaviour observed in the past will be less relevant and attention should be paid to needs that can be expected to remain relatively stable even after big changes. Preferences will become less important than constraints that limit choices.
In this presentation, it will be shown how urban models with different spatial and temporal resolutions can be linked. For this a model system originally developed at TU Dortmund University and applied to many regions and cities from the European to the regional scale will be used as example:

(1) The SASI model is a dynamic simulation model of socio-economic development of NUTS -3 regions in Europe under exogenous assumptions about the economic and demographic development of the European Union and transport infrastructure investments.
(2) The IRPUD model is a dynamic simulation model of intraregional location and mobility decisions of households and firms in medium-resolution zones representing statistical districts connected by a multi-modal transport network.
(3) The RASTER model assesses air quality, noise levels, population exposure, quality of open space and accessibility in high-resolution raster cells as required for modelling noise propagation and air quality.

As an example of a multi-level, multi-scale urban model system a three-level application of the three models for the Dortmund region and ongoing work with the SASI and IRPUD models were presented.

Hydro-Based Metamorphoses in a Polycentric Region: A Multi-Scale Spatial Model for the Ruhrgebiet

Sabrina Sposito

Metropolitan regions portray both the intricately settings and the scales of novel territorial frames that, by embedding a diverging inventory of socio-economic and ecological forces, appear to be undergoing constant change. As a matter of fact, a state of strain manipulates, and then challenges, the urban and cultural spheres of cities and regions (see Brenner 2014; Corner 2006; Corboz 1986; Schröder et al. 2016; Varnelis 2008). In this regard, the triptych of derelict rivers, unstable waters and disrupted cities is barely examined as an intertwined manifestation of certain imbalances and decays, playing at various scales. Rather, such triptych is mostly disassembled into three isolated facets, each a domain of manifold controversies regarding the theory, praxis and politic of space.

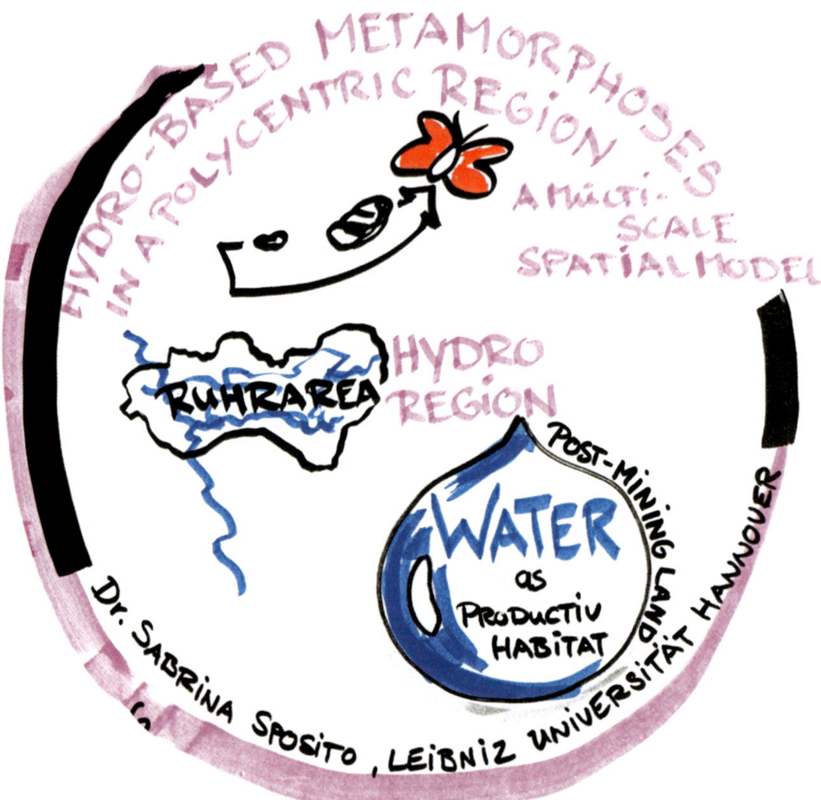

In light of the stated premise, this contribution focuses on a multi-scale spatial model for the Ruhr area: a hydro-region, in which water is at the forefront of experimentations for a renewed regional connectivity. Implemented during a post-doctoral research project funded by a DAAD Research Grant for Doctoral Candidates and Young Academics and Scientists 2017/18, the abovementioned model has addressed the Ruhr area against a broader Northern Europe and German backdrop.

Located in the state of North Rhine-Westphalia and included in the Rhine-Ruhr metropolitan region, for decades the Ruhr area has been thoroughly discussed in relation to the peculiarities of its polycentrism, the legacy of post-mining activities as well as the shrinking patterns arising from a robust structural change (Zimmermann 2017). Ample coal reservoirs, in combination with a transversal inlet of the rivers Ruhr, Emscher and Lippe westwards into the northward Rhine river and a network of channels, used to substantially influence the industrial and urban prosperity of the plain, until its decline in the late twentieth century (Keil & Wetterau 2013). In the aftermath of this process, the volume of the impacts caused by mines and infrastructural operations gained considerable evidence. Thus, since the 90s, the recovery of depleted industrial sites has been tackled in the larger framework of water-related transformations (see MacDonald 1995).

Against a background of both dismissal and regeneration, the recent debate has proved the regional scale as a spatial priority, thus finalising the experiments of territorial coordination initiated at the beginning of the last century (see Keil & Wetterau 2013; Polívka, Reicher & Zöpel 2017; Reicher et al. 2011; Zimmermann 2017).

This contribution formulates the hypothesis that the emergent regional effort in the Ruhr area can serve as a breeding ground for scaling up to a model of hydro-region, in which water is reinvented as the spatial, cultural and productive vehicle for re-assembling the existing urban constellation. This observation presupposes the re-positioning of the valley in terms of various macro and micro systems (navigation/transport, purification/metabolism, production/regeneration and living/adaptation) through the water logics. This study suggests a three-fold method – analytical, explorative and transformative – based on 'retroactive' (Koolhaas 1994) interpretations of the Ruhr urban space and the outlining of (low to extreme) scenarios: forging the infrastructural ground, flooding the subsidence lands, regenerating the archipelago and making the inner islands habitable (Sposito 2018).

Bibliography

Brenner, N., (ed.) 2014: Implosion/Explosion: Towards a Study of Planetary Urbanization. Jovis Verlag GmbH, Berlin

Corboz, A., 1986: 'Il territorio come palinsesto', Casabella 516: 22–27

Corner, J., 2006: ‚Terra fluxus', in Waldheim, C. (ed.). The Landscape Urbanism Reader, Princeton Architectural Press, New York

Keil, A., Wetterau, B., 2013: Metropolis Ruhr. A regional Study of the New Ruhr. Regionalverband Ruhr, Essen

Koolhaas, R., 1994: Delirious New York: a retroactive manifesto for Manhattan. Monacelli Press, New York

MacDonald, R., 1995: 'The Emscher Park IBA', Urban Design 56: 28-31

Polívka, J., Reicher, C., Zöpel, C. (eds.) 2017: Raumstrategien Ruhr 2035+ Konzepte zur Entwicklung der Agglomeration Ruhr. Verlag Kettler, Dortmund

Reicher, C., Kunzmann, K., Polívka, J., Roost, F., Utku, Y., Wegener, M. (eds.) 2011: Schichten einer Region – Kartenstücke zur räumlichen Struktur des Ruhrgebiets, Jovis Verlag GmbH, Berlin

Schröder, J., Carta, M., Ferretti, M., Lino, B. (eds.) 2016: Territories. Rural-urban Strategies, Jovis Verlag GmbH, Berlin

Sposito, S., 2018: 'The Ruhrgebiet – Retrospectives and scenarios of a hydro-region', in Fakultät für Architektur und Landschaft, Leibniz Universität Hannover (ed.) Hochweit 18. Jahrbuch der Fakultät für Architektur und Landschaft. Leibniz Universität Hannover, Jovis Verlag GmbH, Berlin, 165

Varnelis, K., (ed.) 2008: The infrastructural city. Networked ecologies in Los Angeles. Actar, Barcelona

Zimmerman, K., 2017: 'Re-Scaling of Metropolitan Governance in Germany', Raumforsch Raumordn 75:253–263

Polycentric and Multi-Scalar: Technology Transfer from Detroit to Moscow

Claire Zimmerman

As the U.S. and the USSR experienced explosive growth, Detroit and Moscow struggled to catch up with the industrial powerhouses of the U.K., France, and Germany. In one case, a young city expanded radially along belts encircling the core, and then along rays extending out from the Detroit River, creating a network of linear cities aligned with rail systems. In the other, an ancient city centred on a palace complex was surgically altered to accommodate new infrastructure and new industrial installations around its periphery, resulting in similar linear networks cut through older fabric.

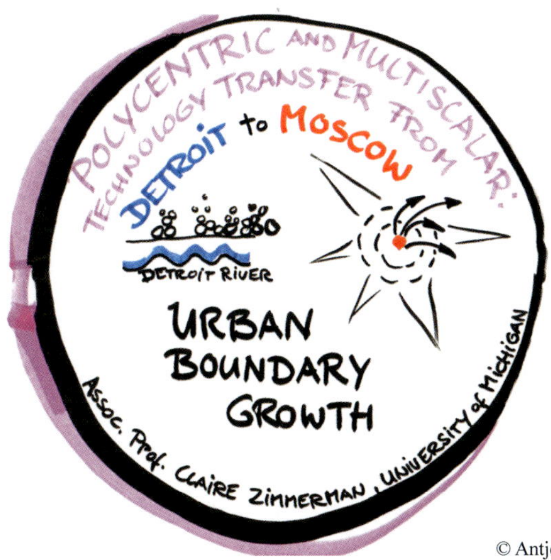

© Antje Rieder, Graphic Recording

Yet Moscow and Detroit were not just cities indelibly marked by the rapid pace of industrialization before WWII. Both headquartered national campaigns of infrastructure construction over a much larger terrain, as epicentres that decentralized industrial installations, accompanying housing settlements and transport networks. The linear urbanism that modernized Moscow and Detroit and the urban nodes that both cities spawned near and far, show how the Second Industrial Revolution generated a newly scaled, distributive urbanism, from Buffalo to Omaha to New Orleans; from Leningrad to Kiev to Baku. Strip development ubiquitous in U.S. cities, and linear urbanism in Stalingrad and Magnitogorsk are here correlated to protocols of urban industrialization, emanating outwards from Moscow and Detroit.

Critical Transformation at the Urban Fringe - Conflict Between National Energy Policy and Local Landscape Value

Hisako Koura

The expansion of using renewable energy sources is the global interest for the sustainable society. The Japan Energy Plan of 2015 presented the projected energy mix at 2030. In this report, the renewable energy share is expected to increase to around 22-24%, which is now 3.2%, and 80% of that is solar energy. The solar energy plants have been planned and developed since the national energy policy pushed their introduction forward. Many of them are located in the abandoned farming fields in the urban fringe and natural environment near the urbanized areas, which cause controversial conflicts with local landscape value. This often occurs due to the different valuation bases in economy and landscaping.

The local landscape value is unique but not special like the cultural assets, which make it difficult to increase social awareness. In Japan, people gradually become conscious of the negative impact of large-scale solar energy plants to the local landscape, and the landscape policy statement by MILT [Ministry of Land, Infrastructure, Transport and Tourism] indicates the necessity of strategic landscape management to cope with other public interests that might damage its value such as the large-scale renewable energy plants. Even though the local landscape value and the renewable energy demand must be considered together and do not contradict each other, it is still necessary to control their locations under the transforming land use conditions to cope with local economy.

Under the present planning system in Japan, we have learned from the recent cases that the following aspects should be discussed to set up an appropriate planning management for these conflicts: 1) the due process of pre-application discussion is necessary, 2) a locational restriction system should be introduced to coexist with the conflicting public interests of the landscape and large-scale renewable energy plants, 3) a support system, which provides expert knowledge and opinion, should be established together with the promotion of local involvement for the local government to confirm their decisions, especially when the development plan cannot be accepted at the projected site.

Smart and Connected Communities Polycentric Development toward the Vision of 21st Century Main Street Communities in Virginia

Mona El Khafif

Cities and communities in the U.S. and around the world are entering a new era of transformational change, in which their inhabitants and the surrounding built and natural environments are increasingly connected by smart technologies, leading to new opportunities for innovation, improved services, and enhanced quality of life. The Smart and Connected Communities (SCC) program supports strongly interdisciplinary, integrative research and research capacity-building activities that will improve understanding of smart and connected communities and lead to discoveries that enable sustainable change to enhance community functioning. The Research Connection Network [RCN] MainStreet21, recently funded by the National Science Foundation and spearheaded by UVA School of Architecture, focuses on medium/small size, remote, and rural communities in the Commonwealth of Virginia through a polycentric (multiple centres) integrated policy, design, and technology approach. The communities served by the RCN have higher barriers to information, resources, and services than larger urban communities.

To reduce this gap, this research project proposes to develop need-based R&D pipelines to select solutions with the highest potential impacts to the communities. Instead of trying to connect under-connected communities to nearby large cities, this proposal aims to develop economic opportunities within the communities themselves. This presentation will introduce the NSF agenda for Smart and Connected Communities and will feature the work conducted so far under the umbrella of the RCN MainStreet21.

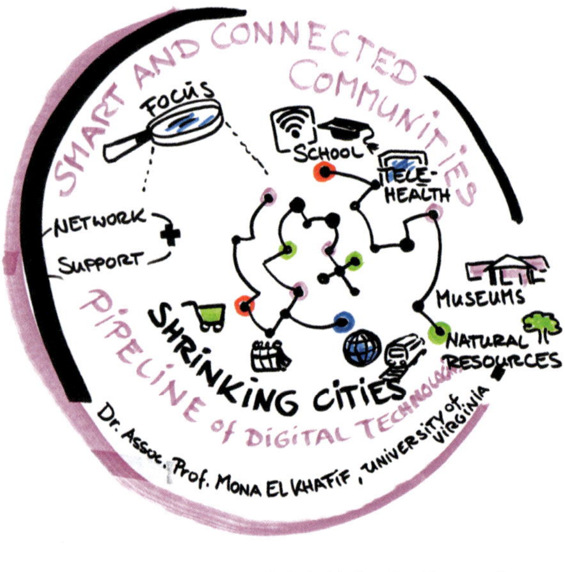

© Antje Rieder, Graphic Recording

2. Resilience + Diversity & Inclusion

Resilience as a Process – A Transformative Concept

Jan Polívka

Resilience is often being understood as a conventional concept describing the power to resist major shocks and changes, as well as the ability of a system to restore itself after such changes. While this understanding remains relevant, the current resilience approaches merely emphasize the adaptive potential of resilient systems, thus pointing out their power to develop, to adjust and to reorganize. Not only the transformation helps systems to get back to where they were; it also enables them to willingly change their functional constellations. To be able to do so, systems interact with and creatively absorb new influences. They also reorganize themselves internally and acquire and process an ever changing variety of resources. In doing so, resilience enables the quite complex and resistant human-nature nexus of our living environment to even radically change in order to remain existent. Not all such ways are stable and solid, safe in result, formal or even legal, but most of them remain highly innovative in process and edge breaking in results. The keynote calls for both urban/regional experiments and the space for spontaneous irregularities as an integrative trial mode to re-define our settlement space, enable it to develop, and thus allow it to survive.

© Antje Rieder, Graphic Recording

The Geographies of Urban Resilience in German Large Cities

Asad Asadzadeh, Theo Kötter, Dominik Weiß

Motivation and target

A series of prominent natural and technical hazards (hurricanes Katrina and Sandy, the Great Eastern Japan, Haiti and Nepal earthquakes), as well as socio-economic and socio-political shocks (civil war in Syria and the entrance of 1 million refugees to Germany in 2015) highlight that our communities are not immune to the forces of sudden changes. Since cities are at the forefront of these challenges and their consequences, international academic and policy circles have acknowledged the need for making safe and resilient cities (UNISDR 2015; UNSDG 2015; UN-Habitat III 2016). Resilience encompasses the way in which cities face the adverse events and consists of inherent and adaptive capacities of them to respond, adapt, and grow no matter what kind of risks they experience (Rockefeller 2015). Since

© Antje Rieder, Graphic Recording

resilience is a place-specific and multi-dimensional term, operationalizing the concept can predispose decision-makers, stakeholders, and urban planners to understand factors that contribute to resilience and interventions that are needed to build and enhance it (Asadzadeh et al. 2017). Therefore, the project entitled "Stresstest Stadt" aimed at operationalizing the concept of urban resilience among German large cities and provided a benchmarking tool for assessing resilience level within the study areas.

Theory and method

Operationalizing urban resilience is primarily started by asking the obvious questions of resilience of what to what. The former (what resilience) focuses on the semantic completeness of resilience measures and denotes whether measuring a set of attributes or assets (result-oriented) or assessing a set of capacities and processes (process-oriented) within cities or both. This study views urban resilience as a dynamic concept and considers it as both the static (robustness) results and dynamic processes (adaptability).

The latter question (resilience to what) matters to the development of underlying assumptions and descriptors of resilience frameworks that are place-specific and multi-scalar. This is because urban strategies and interventions may vary depending on which stresses and shocks are considered.

This study has focused on those probable challenges which are expected to occur over time in German large cities. For this study, therefore, eight major stress scenarios were developed that can have negative effects on main urban functions such as demographics, infrastructure, environment, etc. To understand whether and to what extent a city will be able to maintain its functionality in time of an adverse event, those eight descriptors were selected which represent the different sectors of a city to fulfil the basic urban function.

The developed stress-scenarios and respective descriptors are the result of a consensus-based discussion process in the advisory board of the research project. For each stress scenario, those descriptors are considered that can be used to determine the resilience level of a city, defined as robustness and adaptability. The robustness dimension is measured by quantitative statistic data and can be presented in a nationwide ranking of all cities. The dimension of adaptability is a qualitative audit among experts of the city administration. The questionnaire evaluates governance concepts, institutions and strategies that help cities to respond and to adapt properly concerning the examined stress-scenarios.

Empirical results

The empirical application of the stress-scenarios was carried out among 106 German large cities. The results provided a resilience profile for each city in the form of a matrix that displays the robustness and adaptability of the stress scenario(s) .

The determination of the resilience profile provided an integrative and comparative result of the stress test against the eight potential challenges. Using this classification scheme, the representation of the robustness enables a ranking within the range of the urban municipalities in Germany. It can be determined in percent how many cities have a higher or lower assessment of their robustness in a specific stress scenario and the relevantly affected descriptors. The developed framework and its methodology propose a flexible and transparent idea of why resilience assessment is being performed and what the assessment intends to achieve.

Bibliography

Asadzadeh, A., et al., 2017: Operationalizing a concept: The systematic review of composite indicator building for measuring community disaster resilience. International Journal of Disaster Risk Reduction, 25, pp.147–162

Rockefeller, A., 2015: City Resilience Index, Available at: http://publications.arup.com/Publications/C/City_Resilience_Framework.aspx

UN-Habitat III, 2016: New Urban Agenda, Available at: http://habitat3.org/wp-content/uploads/New-Urban-Agenda-GA-Adopted-68th-Plenary-N1646655-E.pdf

UNISDR, 2015: Sendai Framework for Disaster Risk Reduction 2015 - 2030, Available at: http://www.preventionweb.net/files/43291_sendaiframeworkfordrren.pdf

UNSDG, 2015: Transforming our world: the 2030 Agenda for Sustainable Development, Available at: http://www.un.org/ga/search/view_doc.asp?symbol=A/RES/70/1&Lang=E

Resilience Trade-offs: A Dortmund Tale of Tribulations with Informal Initiatives and Inclusionary Processes

Robin Chang, Silja Kessler, Christian Lamker

Resilience gains support in academic scholarship and public policy with its focus on (urban) systems by advancing reactive and proactive responses to external opportunities and shocks, but also increasingly by pointing towards different insights for open and inclusive planning. This contribution presents current contentions between developing inherent adaptive capacity at the neighbourhood level and finding entry points to include multi-cultural community citizens in externally initiated processes of urban planning in the city of Dortmund. We take an evolutionary perspective on the performance of social-ecological resilience for inclusive local transformations embedded in the broader fabric of post-industrial developments. Specifically, informal networks emerging around the re-use of an old industrial site west of the city centre that was formerly used for sheet pile wall construction (Hoesch Spundwand und Profil GmbH) are narrated.

The term of resilience has been framed in many ways with regard to public policy and planning, especially along conflicts between keeping the status quo to changing the whole system: Handmer/Dovers (1996) define three types of resistance/maintenance, change at the margins and openness/adaptability, whereas Hillier (2015) uses the terms of persistence/resistance, adaptability and transformability. We acknowledge unavoidable trade-offs between open and transformative processes and institutional and spatial realities. We outline positive potentials of resilience as a concept to plan in open and inclusive processes that add to the transformation of post-industrial regions. We background our presentation with an understanding of resilience as connected to inclusive planning, engaging with local communities and potentially transforming urban areas and regions.

Therefore, we ask:

(1) To what extent does resilience scholarship help in considering evolutionary processes unfolding between professional planning, property owners and emergent civil society initiatives?

(2) How can urban researchers refine conceptual resilience in urban planning theory and practice to develop social and cultural astuteness and overcome engagement barriers?

The contemporary struggles of civil society to engage with property owners, public administration and ordinary people for access and use of vacant industrial land set the tone for a participatory process that volunteers along with Die Urbanisten, a local NGO, are trying to establish. In the midst of establishing and negotiating participatory legitimacy in the name of adaptation, community volunteers forming a group Neue Werk Union are confronted with social trade-offs such as ambivalence and consequential exclusion. Actual examples with strong property owners, vested economic interests and large potentials for conflict challenge existing community vulnerabilities. Calls for self-organization and capacity building in light of community and social resilience in urban neighbourhoods are seemingly at odds with such post-industrial realities. Manifold ideas for temporary and permanent uses and for designing a process to get to an agreement confront established ways how to develop an old industrial site to a new inner-city suburb.

We see the potential for transforming institutions and finding solutions for co-creation and experimentation by taking a different conceptual lens that puts a stronger focus on citizen-driven urban development (Bisschops/Beunen 2018). We put forward the power of openness, adaptability and inclusion to foster transformative actions towards emergent joint futures that keep a balance between stability and flexibility. This perspective opens ways to address simultaneous disagreements about goals and a perceived lack of knowledge and, therefore, overcomes barriers caused by available knowledge, or construction thereof by powerful actors (Bertolini 2010; Hillier 2015).

Based on auto-ethnographic experiences, we explore and discuss the intersection between informal intervention processes and transformative collective (inter) action in the context of post-industrial developments. Informal and temporary urbanism and evolutionary economic geography help provide explanatory value in the struggle for social-relational and subversive resilience.

Bibliography

Bertolini, L., 2010: Complex Systems, Evolutionary Planning? In de Roo, G. &. Silva, E.A. (eds.), A planner‘s encounter with complexity (S. 81–98). Burlington, VT: Ashgate

Bisschops, S. & Beunen, R., 2018: A new role for citizens' initiatives: The difficulties in co-creating institutional change in urban planning. Journal of Environmental Planning and Management, 15(1): 1–16. doi: 10.1080/09640568.2018.1436532

Handmer, J. W. & Dovers, S. R., 1996: A Typology of Resilience: Rethinking Institutions for Sustainable Development. Organization & Environment, 9(4): 482–511. doi: 10.1177/108602669600900403

Hillier, J., 2015: Performances and Performativities of Resilience. In van Assche, K., Beunen, R., & Duineveld, M. (eds.), Evolutionary governance theory: Theory and application (S. 167–183). Cham, New York: Springer

Socially Built Housing and its Impact in the Development of Inclusive Neighbourhoods: The Latin American Case

Marielly Casanova

Social housing in Latin America can become a model for the implementation of housing solutions and integration for refugees in the European context, especially in Germany with the mass migration during the last few years.

Housing is a very sensitive topic, especially in the context of refugees and disadvantaged communities. Housing arrangements represent a powerful mean that defines the future of migrants and their opportunities for integration in the society. Therefore, the current discussion surrounding the sustainable character of housing and cities should be directed towards social strategies and their potential to build affordable housing, inclusive neighbourhoods and the development of a public policy for social and economic development. On the contrary, wrong approaches resulting in homogenization (lack of roots and identity), lack of participation in the process (no decision-making) and disconnection from urban life are the fundamental causes for exclusion, decline of neighbourhoods and stigmatization of the inhabitants, impairing their opportunities to integrate both socially and culturally into their environment as well as in the economy. The cost of recovering ghettos and marginalized communities is even higher in the long term than investing in comprehensive urban strategies to build cohesive communities today.

Moreover, in general approaches to urban renewal, rehabilitation and the construction of low-income housing, the social and cultural dimensions are often neglected. Then, how socially built housing could lead into neighbourhood regeneration? How to develop vibrant and healthy neighbourhoods with social housing as a means for urban development and social integration?

© Antje Rieder, Graphic Recording

Social initiatives, organized communities and empowered groups in Latin American cases have shown the enormous regeneration potential that they represent. Self-organized and self-managed groups, based on cooperation principles, have been transforming their habitat, economy and quality of life, while paving the way towards a new social housing methodology. In this sense, social housing or socially built housing is interpreted as a complex system integrated by social, economic, political and city making processes. Social practices in the so called "social production of the habitat" provide clues to understand an alternative way of approaching housing solutions in which several dimensions coexist. The social dimension ensures the creation and strengthening of social capital through self-management and mutual-aid. Social capital at the same time is the platform for the development of a social economy based on collective effort and collective benefit. When the habitat is shaped or created by social actors who share interests and needs it becomes a complex process of management and construction of infrastructure - including housing and the surrounding neighbourhoods – as well as the construction of cultural identities, economic process and political decision-making.

Even though the Latin American context is very particular, the same principles could also apply in other regions. By understanding the character of each small fragment or the neighbourhood level, a more inclusionary governance can be established. The same could enable active participation or co-creation by leveraging on the capacities of the population. Working for the consolidation of small scale organizations and initiatives and articulating them with the production of their habitat, along with city planning strategies, could ensure more effective processes for the continuous regeneration of the social, economic and cultural fabrics in different contexts.

The topic of housing is very complex to analyze, and there is an array of elements that need to be articulated. Therefore, through the rationalization of social (self-management), economic (social economy) and urban (neighbourhood making) principles, typologies were constructed for the documentation and evaluation of several case studies in Latin America. At the same time those typologies have served as tools to analyze and evaluate different projects for refugee housing – both temporary and permanent – in the Ruhr region. The contribution from this work could represent a foundation for future research and conception of social housing policies and programs in different contexts including Europe where in later years economic, urban and representation crisis have emerged as the result of exclusive politics, foreign wars and new waves of population.

From Ruins to Revitalization: A Comparison of Regional Strategies and Outcomes

Scott Campbell

A region's historical legacy of heavy industrial production does not necessarily condemn it to post-industrial decline and decay. 20th century manufacturing generated generous profits and relatively high wages, fuelled mass-scaled housing construction and the rapid expansion of the middle class and its institutions. And many cities were able to convert their decades of industrial growth into post-industrial revitalization and reinvention, and renovate crumbling industrial landscapes into housing, offices and public spaces with an emerging post-industrial aesthetic (as seen in the Ruhrgebiet and elsewhere). But the tragic paradox of Detroit is how so much concentrated industrial capital led to so little regional economic innovation after the assembly lines and supplier chains shut down. It's tempting to employ the anthropomorphic metaphor of resilience as an explanation, but diagnosing Detroit's decline as a "lack of resilience" is a better description of symptoms than underlying causes.

This paper explores the competing narratives that divide the struggling and thriving older industrial regions. The first narrative is perhaps the oldest: thriving regions developed diversified economies with rich vertical and horizontal networks, while lagging regions were overly dependent on a single sector that ossified skills, capital and innovation — and crowded out alternative sectors. A second narrative is the role of growth coalitions: strong and strategically unified in revitalizing regions, weak and disconnected in lagging regions. A third lies in the interaction between growth and development: strong regions continually convert industrial profits into locally clustered R&D activities, while weak regions squander or leak profits and fail to plant the economic seeds for the next generation of product- and process-innovation. A fourth lies in metropolitan spatial patterns: the simultaneous development of well-linked suburban networks and a revitalized urban core leads to stronger regions, while the lopsided growth of one at the expense of the other weakens the region. And then Detroit presents its own distinctive paradox: the intense early 20th century concentration of industrial production in one city created the very technology (the car) that would accelerate the flight of people and jobs from the industrial city to the suburbs, and from the Midwest to the growing American South and West. And all of this was (and remains) exacerbated by the unresolved racial divide between city and suburb.

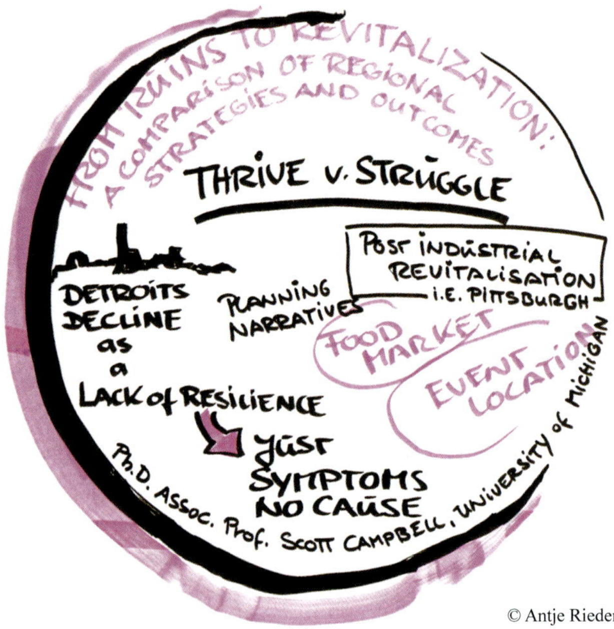

© Antje Rieder, Graphic Recording

All these explanations are endogenous: linking the fate of the region to internal characteristics of its land use, industrial mix, culture and infrastructure. But the contrast of Detroit and the Ruhrgebiet highlights another, external narrative: do larger governmental authorities (states/Länder or the federal level) step in with a strong, well-funded system of redistributive support, or does the state turn its back on the city in a national culture of anti-urban policy?

This paper concludes by contrasting the key elements of 20th century vibrant industrial regions with those of contemporary post-industrial regions. Regional development advantages were once built on river, canal and rail lines, clusters of ports and warehouses, and a strategic location at the midpoint between backward linkages (natural resources and cheap labour) and forward linkages (markets for industrial goods). The contemporary region has a new political economic geography, with new infrastructural and land use patterns. Footloose labour and capital lead to the competitive attraction of highly-educated "talent" in this era of amenity-driven development. Clusters form around universities and research centres, airports, transit and highways. New forms of work and new networks of commuting lead to unexpected geographies of labour markets. And the recent announcement by Amazon that Detroit did not even make the list of top 20 finalists in its search for a second 50k employee-headquarters (HQ2) sent a clear message to the region that Michigan had not yet navigated the pathway to the new economy.

Heritage Planning for a Polycentric City - Historic Concepts and Ideas for the Future

Carmen M. Enss

The research group for urban conservation at the University of Bamberg, Centre for Heritage Conservation Studies and Technologies (KDWT), studies historic concepts and develops new methods for integrating urban planning, participation, and heritage conservation. One goal is to connect heritage conservation to city planning within a joint process. We theorize heritage concepts for polycentric cities and regions. The focus is not only directed to historic city centres (Altstädte), but towards new and integrative concepts of "Heimat" and heritage. While established concepts of heritage are rooted in a national character of architecture, new planning tools need to re-connect urban heritage to diverse social groups or inhabitants.

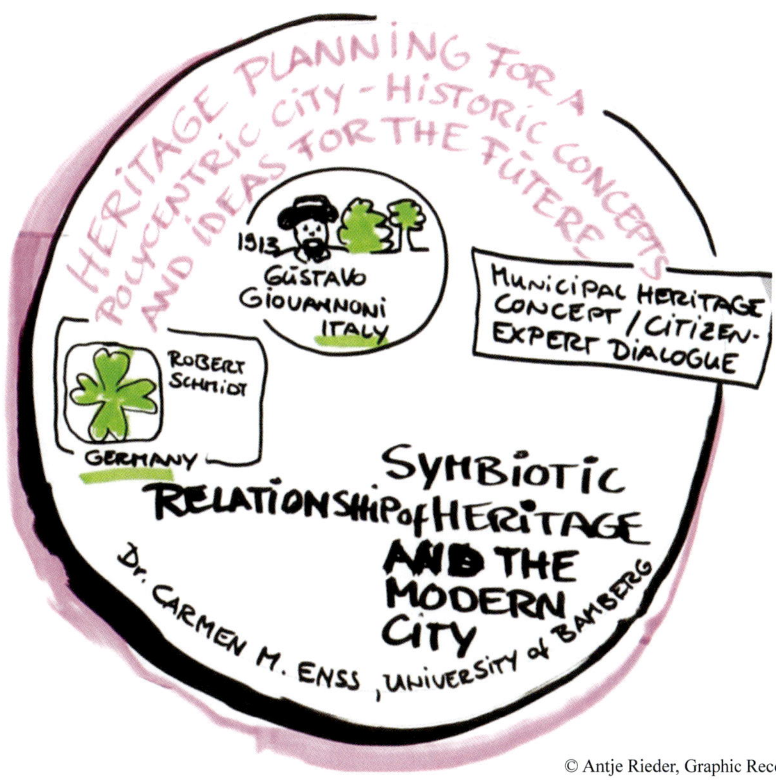

© Antje Rieder, Graphic Recording

The conference paper analyzes historic model designs that combined heritage planning with planning for city transformation processes. Strategic development plans for the historic urban fabric were proposed as early as 1913 by Gustavo Giovannoni, an Italian city planner and conservation theorist. Giovannoni formed a part of the international garden city movement. His Italian architecture school, which was based in Rome, applied garden city concepts not only to city extension areas, but to the city centres as well. This was done in order to improve living and working conditions in dense urban areas, but, what's more, for heritage conservation reasons. Giovannoni's garden city Aniene near Rome shows in different scales some similar structures as were realized in his plans for Rome's renaissance quarter in the historic centre. Giovannoni's concept of urban heritage areas culminated in the heart of the city. However, he also included large park areas and archaeological zones into his ideas of an urban heritage. Concurrently, in Germany first systematic regional planning efforts were made with the development plan for the Ruhr region from 1912 (Generalsiedlungsplan Emscher-Ruhr). While the Bauhaus school of architecture focussed on social aspects in city planning since 1919, Giovannoni's school in Rome continued to focus on a national style and heritage aspects in planning, often omitting social aspirations in urban planning.

Later in the 20th century, regional and heritage planning ideas in Italy led in combination to a concentration of conservation practices with heritage zones like old towns. Most innovative in their times, Giovannoni's model designs were mostly realized within the "walled city", leading to homogeneous heritage zones which formed distinctive parts of the urban fabric. Heritage zones became part of modern functional city planning (Gerhard Vinken: Zone Heimat 2012). Our effort is to overcome heritage planning for closed zones and to integrate heritage aspects within the city planning process. We can use Giovannoni's ideas to combine city planning, architecture and conservation planning in all areas of the city, especially to post-war cities. But they need to be broadened to include more features: Social planning and diversity planning aspects can inform new developments in heritage in order to provide contemporary bases for integrated cities. The paper suggests links from historic heritage planning principles to nowadays heritage and diversity planning for strengthening the identity of – and the connectivity between – urban quarters and neighbourhoods.

Collective Challenges and Joint Arrangements in the Dispersed Territory of Flanders

Maarten Gheysen, Kris Scheerlinck, Erik Van Daele

This abstract builds upon three conceptual lines. These lines are brought together in a central statement that was dealt with during a research by an international design summer school and the design studio at the Faculty of Architecture, KU Leuven (Belgium).

Conceptual line 1: The dispersed territory is an urban system

Over the last decades, the urban paradigm underwent an enormous shift. The traditional dichotomy between urban/rural has altered as the result of a continuous process of growth and (re)distribution of parts of the city. As a result, the relation between the new territories of dispersion and the traditional city is no longer a centre-periphery one, but that of dialogue between two types of the city. Often considered the antipode of the compact city, the diffuse city became the object of research, leading to different interpretation, but all pointing to "a kind of hybrid megalopolitan spatial organization characterized by the presence of certain urban characteristics in the absence of other"(Vigano, 2012).

© Antje Rieder, Graphic Recording

Flanders, as part of the larger North-Western Metropolitan Area, is no exception to this paradigm shift. It is a project we started to call All City/All Land (AC/AL). However, as indicated by many authors, despite the acknowledgement of this paradigm shift, an adapted design vocabulary and theoretical framework is still missing (Segal and Verbakel, 2008).

Conceptual line 2: Collective challenges

Despite its qualitative living conditions, omni presence of landscapes, resilient structures, and a proximity of living and working, several collective challenges (collective since they cannot be tackled on the individual level) such as "environment, mobility and the growing social inequalities" (Secchi, 2009), are being raised in regard to AC/AL, and for which the dispersed city is being held responsible. Dispersion is often considered "bad", while the compact city is thought to be the only "good" solution, thereby neglecting the dispersed reality in which more than half of Flanders' inhabitants work and live (Vermeiren et al., 2018). But as Secchi (idem) stated, it is exactly when these major collective challenges coincide and are being addressed, a new form for the city appears. In other words, the collective challenges offer the opportunity to update the design vocabulary and theoretical framework.

Conceptual line 3: Joint arrangements and positive externalities

The last conceptual line focusses on an economic concept called 'positive externalities' as formulated by Arthur Pigou in the 1920s; namely '… the positive effect an activity imposes on an unrelated third party.' (Varian, 2010).

In 1858, the city of London was confronted with a major collective challenge, the Great Stink, linked with contagious diseases, such as cholera. To cope with this, a major sanitation was set-up, constructing a large-scale sewage system and the embankment of the river Thames. Literally on top of all these infrastructures, new streets and parks were built. The construction of these public spaces was never the central goal of the project, of the joint arrangement, but can be interpreted as a 'positive externality'. It means that tackling the collective challenges of the dispersed territory by joint arrangements can generate positive externalities.

Research statement

The increasing complexity of our society confronts our (urban) systems, in particular AC/AL, with challenges of a collective nature far beyond the complexity of the Great Stink. These challenges have partially been tackled over the last years, mainly related to large sewage programs, traffic safety and attention programmes to disadvantaged demographic groups. As a positive externality we saw the refurbishment of many village squares and roads. However, the design vocabulary of these refurbishments is not adapted to the conditions of AC/AL and expresses a spatial dyslexia and agraphia (Gheysen, Scheerlinck and Van Daele, 2017). Still, by addressing these collective challenges in the dispersed city, a new form for the territory can be made through positive externalities.

Designerly research

To test and illustrate this research statement, an international summer school and subsequent design studio were held in 2017, elaborating the availability of water, mobility or the access to resources, the ageing of neighbourhoods, and 3rd generation energy landscapes as themes. The underlying design question was not to design an energy network nor a water system, but was found in the question what positive externalities could arise when AC/AL is undergoing vast investments and changes.

In general, the following conclusions were made:
The logic of water and energy let them touch upon different parts of AC/AL, opening up the possibility to generate joint arrangements in the whole territory through landscape or infrastructure supported interventions. Mobility, in contrary, is a punctual type of intervention, much more operating at specific points where flows are compressed, a joint arrangement as singular intervention.
To fully grasp the opportunities of the positive externalities one cannot rely on the vocabulary of the compact city. AC/AL requires an attitude of creating opportunities rather than results, using open forms, diversity, and complexity as language, in relation with (natural) processes, and by doing so confirming the hybrid character of AC/AL.

Bibliography

Gheysen, M., Scheerlinck, K. and Van Daele, E., 2017: 'Opposing Oppositions, All City/All Land', MONU: magazine on urbanism, (26), pp. 62–67
Secchi, B., 2009: 'Une nouvelle question urbaine', in Colloque internationale. Le territoire dans tous ses états. Paris: SFA
Segal, R. and Verbakel, E., 2008: 'Urbanism Without Density', Architectural Design, pp. 6–11. doi: 10.1002/ad.601
Varian, H. R., 2010: Intermediate microeconomics: a modern approach. W.W. Norton & Co., New York
Vermeiren, K. et al., 2018: 'What is Urban Sprawl in Flanders?', in Expanding cities, diminishing space. Vienna, pp. 537–545
Vigano, P., 2012: 'The contemporary european urban project', in Crysler, G., Cairns, S., and Heynen, H. (eds.) The Sage handbook of Architectural Theory. Sage Publications, pp. 658–670

Resilient Land Use Development for Illegally Developed Areas: The Kyoto Fushimi Greenbelt

Kiyoko Kanki

Where and what happened?

At the southern foot of the famous Fushimi-Inari shrine hill, known as Inariyama, we can find an uncoordinatedly mixed land use of industry, waste dumping, housing, vacant or green open space without standard infrastructure. These areas are illegally developed areas, which had started just before - and continued even after - the land use zonings were introduced in Kyoto City in 1971, and illegal incineration continued till the late 1990's. These areas had been the historic soil production sites necessary for the traditional walls and roofs of Kyoto's temples, shrines, as well as for houses. At the eastern part of the Kyoto greenbelt, it is important to find a practical way for these areas to be re-designed into legalized land uses and preferable landscapes.

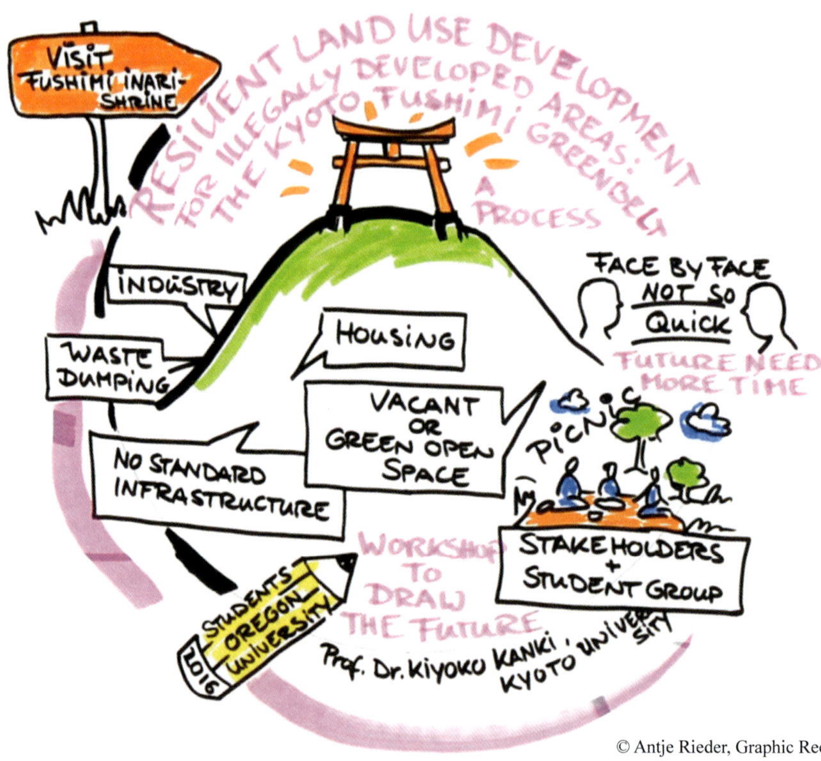

© Antje Rieder, Graphic Recording

Problems

In 2010, the city government decided its policy for these areas to promote legalization and to introduce district planning methods for ordering land use and improve environment. Indeed, there was almost no idea how the land use should be like here, and how to foster the transition into a preferable land use. Stakeholders on the sites are generally low capacitated for new self-investment, and they had not been well coordinated because they themselves were recently under pressure of illegalness. In several sites, public authorities forced the legalization of structures by defining a certain size of roofs, walls, or beams with no permission had been operated by the city's central government until 2 years ago. In tandem, the Fukakusa branch office of the city government was responsible for the preparations of the district planning, and Fukakusa branch officers had to start developing the relationship with each stakeholder to have a trustful communication. At this moment, we (Kyoto University) began consultations with the Fukakusa branch to find an innovative idea about the development for these areas.

Who can draw the future? – The 3rd group

Direct communication about the future of the areas between the Fukakusa branch office and the stakeholders has not been easy even after they became able to talk rather frequently about the daily news. In 2016, the student group of the Department of Policy Science, Ryukoku University, which is located nearby the areas, started monthly (sometime weekly) activities by having small picnics, inviting the stakeholders, interviewing, and enjoying life together in cooperation with the Fukakusa branch. This activity was very successful and is being continued until now. The atmosphere when some visitors walk through the areas, is rather friendly now, and it became easier to meet and talk with the people of the areas. The students as the 3rd party did change the situation.

International interest and cooperation

Since 2016, the areas have attracted wider interest. In 2016, student groups of Oregon University held a workshop to draw the future image of the areas freely. In 2017, TU Dortmund University set up a more collaborative workshop for professional planning solutions. These opportunities can be regarded as suitable promoters because it is difficult to draw the first idea of planning within the locals. These workshops suggested various ideas for the areas not only focusing on the district planning. Several sites were recommended to be the first step of landscape upgrading to produce comfortable open spaces for stakeholders as well as for visitors.

Starting with new open spaces

After the 2017 workshop, the Fukakusa branch office itself began to re-engineer the landscape, utilizing city owned land. This was turned into new green open spaces with a small river catching the rainwater from Inariyama, and includes a low hill with an old small shrine. In early 2018, the new green open spaces were almost ready for use. The Kyoto University team has followed up the result of the 2017 workshop with GIS analysis of the land ownership distribution and concluded what types of the organization of stakeholders and related locals were necessary to realize the attractive ideas of the 2017 workshop. To realize the wider-scale ideas' essential in the future we need more time to establish a well-connected organization. This showed us that starting with the utilization of city owned property might be the best way, and that starting with expanding the comfortable open space might be suitable and possible to motivate the stakeholders to image the possible upgraded environment of their areas.

Next step

Recently, more tourists like to visit the Fushimi Inari Shrine, walk through the Inariyama and come through the areas. Most of the tourists lose their way to the nearest railway station in these areas, so it became suddenly important to put up guiding information. This is a new chance to motivate all parties involved. This year, the next step for imaging the future will start at the new green open spaces. The 2017 workshop was the trigger.

Reconnecting People with Nature. Green and Social Infrastructure for Urban Integration

Andreas Kipar

> *"A strange spirit of community is conquering the streets and squares, new forms of interaction are being created (...). Many of the urbanist movements set a pragmatism of sympathy and sharing against the economical nature of selfish hearts. The city is dead, long live the city!"*
>
> Hanno Rauterberg, We are the city! (Suhrkamp, Berlin 2013)

We live in revolutionary times. Two hundred years after the Industrial Revolution, the digital revolution is turning our societies upside down, changing the way we think and act. One visible characteristic of digital modernity is networking. But networking what? And how?

In the 19th century, the Industrial Revolution also triggered urban turmoil. The step away from being a fortified city, spreading beyond the medieval city walls, the cutting of embankments, was a step into freedom, a step out into the countryside. But the expansion of the urban world typically took place at the expense of the natural world. The future of the city continued to be defined by structures; nature remained a decoration or at best was permitted to express itself as a designed park, like an island in a stream of buildings and streets. Meanwhile the surrounding countryside became a sort of demi-city, and the growing cities themselves threatened to become peripheral.

The green decade

Ten exciting years lie ahead of us, a "Green Decade". There are seven green corridors in the Ruhr area, between the Ruhr, Emscher and Lippe rivers, as existing structures and cultural inheritance. The residential development process will be completely redefined, following an ecological footprint by which the 53 towns will be positioned appropriately, and then new projects will blossom, allowing green infrastructure to grow.

2027 and beyond: "2027 plus" is a Green Decade that lies open to the future. Fortunately, projects networked with the countryside are also being planned elsewhere. Munich is heading for "Deceleration, densification, transformation" in its Open Space Concept for 2030. Berlin is developing a strategy for urban landscapes: "Natural, urban, productive". In issuing a new edition of STEP 2035,

Vienna is offensively thinking about an efficient green and open spaces concept. Everyone is looking far beyond 2030. In the European context, the northern countries are the pioneers. But even metropolitan regions such as Paris, London and Milan have now started down this road.

The challenges we face are climate change, new forms of living together, and ultimately how to increase the performance of our urban systems. The European Union calls this "nature-based solutions". The Ruhr metropolis can act as a catalyst here. It is already in the middle of a process, and symbolizes this as a blueprint for a tangible green future. The city is not dead, finding itself again in green networks and coming back to life.

© Antje Rieder, Graphic Recording

3. Integrated Urban Regions: Academia, Education, Practice

Exploring Together and Learning from Each Other

The plenary session "Integrated Urban Regions: Academia, Education, Practice - Exploring together and learning from each other" concluded the international symposium "Urban Integration – From Walled City to Integrated City." Prof. Christa Reicher (TU Dortmund University) and Prof. Dr. Angela Million (TU Berlin) moderated a lively conversation with Prof. Jonathan Massey (Dean, Taubman College), Ila Berman (Dean and Edward Elson Prof., University of Virginia), Dr. Uli Paetzel (Emschergenossenschaft), Dr. Hadia Straub (RVR Regional Association Ruhr), Prof. Dr. Hisako Koura (Kobe Design University), and Prof. Dr. Stefan Berger (Ruhr-Universität Bochum).

© Antje Rieder, Graphic Recording

Reicher to partners from practice:
Why is it interesting for you to participate in such a discussion?

Straub:
It is important to be informed about new developments and to get insights from different perspectives.
We, the RVR (Ruhr Regional Association), take part in an exchange of ideas with scientists and experts, and foster knowledge transfer between science and politics. We try to act like a mediator and think tank.

Paetzel:
We try to cooperate with planners, municipalities, politicians, (…), and especially with scientists. It is important to us to have a co-operation with universities and young people, to get new ideas and skills, also from abroad, so that we can improve.

Reicher to Berger:
What is the interest for a historian in this congress?

Berger:
If we are interested in understanding former regions of heavy industry or big urban conglomerations, it is incredibly important that we need to get away from disciplinary boundaries. The institutional kind of disciplines is finally invaded. So we have to familiarize with perspectives that come from other disciplines; that means for me as a historian, to look from the perspective of a planner. In a historical research project at Ruhr Uni Bochum, we also benefit from the inputs of sociologists, of history specialists, of philosophers, of political scientists and even of city planners.

Million to experts from academia:
How can scales between local and regional be connected through urban design and urban planning actions? In which spaces and places should we start with multi-scale strategies?

Massey:
The city planners of Detroit have been developing some very innovative new strategies trying to leverage private investments happening in the city, to spread the benefits and to share the prosperity more equally. They work in collaboration with us (the architecture faculty at Taubman College). Our students are testing scenarios for designing and developing missing middle-rise housing. This can

show the developers and neighbourhood residents what it might look like if the plan were to become operational. They are working at the scale of a very small housing project but linked to a citywide plan for housing.

Berman:
On the one hand, academic institutions take on with issues and problems. But on the other hand, within the private and often the public world, there are not even the time or the resources to deal with these things. So we take the intellectual capital, the design capacity and innovative capacity of the institution in order to apply it to the room every day. One of the challenges is often organizational, to deal with the multiple-scales and at the same time to be able to work across disciplines. Disciplinary expertise and specialization are necessary. But the challenge is to create the ‚We', to be able to deal truly with multi-disciplinary problems and to be able to work across large-scale projects that operate over a long period of time, with collaborations that can work across disciplines.
There are different developed strategies, but we have not really landed on the most productive one. The challenges that we have are so large that we are going to require large-scale collaborative efforts and the kind of organizational structures.

Million to Koura:
How was it in Japan?

Koura:
In Japan, most of the planners are educated in other backgrounds than in planning in the narrow sense. In the 1960s, when expanding of cities started, academia, the public sectors and planners worked together and the profession of urban planning developed. But simultaneously with the shrinking of the cities, when the strategic planning approach is needed for delivering sustainable transformation, the planning division is also shrinking.

Reicher:
How can we organize a kind of knowledge transfer? Must there be an institutional transfer?

Berger:
There is knowledge transfer happening at a lot of levels, but we do not have this sensitivity to local conditions, for particular historical, social or political contextualization. Taking something that works somewhere and transplanting it somewhere else and thinking surely it works there as well, does not work. It is still quite rare to have people who have this kind of in-depth knowledge.

Massey:
In San Francisco, California, there is a year-long research and design process called ‚Resilient by design – Bay Area Challenge‘ which is wired by rebuilding projects in New York Bay after Hurricane ‚Sandy‘. The initial aim was to transfer knowledge from New York to San Francisco in order to have academics and practitioners teach policy makers and citizens how to plan comprehensively around resilience. At one level, practice and academia learned from one another, but also learned from the citizens in some neighbourhoods who pushed back against top-down governance.

Reicher to Berman:
You established this kind of ‚Next Cities‘ institute. Is this an institution which can organize knowledge transfer?

Berman:
It was meant to bring together a number of different projects that were happening all over the world. The idea of the institute is to operate with the universities and partner institutions, but also rather than always reinventing models that we could look at how best-practice is implemented. It is about how we would essentially

develop forms of knowledge transfer from one kind of project or region to another. There are issues we are dealing with, that are similar and can occur between one environment and another.
Even among our own internal disciplines we do not have enough knowledge transfer.

Straub:
Knowledge transfer is always about individuals and communication – that's why the Ruhr Regional Association likes to participate in such conferences. Experts among their own discipline and especially from different disciplines need networking and communication. That can be a big challenge in an accelerating working environment. Here, communication sometimes is seen as a hindrance, especially communication beyond the boundaries of the own discipline where the outcome is uncertain at the beginning. As an administration, we at the Ruhr Regional Association regularly face the challenge to communicate professional results of let's say planners to the expert sphere of political decision-makers. Another aspect of knowledge transfer is the need for people to continuously deal with a single topic. This is important in order to use the tacit, undocumented knowledge they gather over time. This continuity gets lost in a working environment where experts are hired on a project basis or as consultants.

Koura:
It is also important to get the knowledge of the local community. There must be many different scales of approaches and methods to share knowledge even for the site-specific and local events, and appropriate interpretation of knowledge among community, professional planners and provisions is essential.

Million:
Do you have to live in a metropolitan region, to be part of the development and the design? I always use the example of: if you do not know how to ride a bicycle, can you really plan a bicycle road system?

Berman:
You can ask the question in other disciplines, too. Do I have to have a faulty heart to be able to do a heart transplantation? In my project for example, the students do not have the experiences and have not ever been to India. In the process they are doing research from distance, but also in India. There is something from not being close to a problem because when we come from a particular area, we have heard all of the solutions and we cannot imagine a new way of seeing problems. So maybe we do need somebody who has not ridden a bicycle, to actually rethink the structure of the bicycle.

We need those who are looking at it from an entirely different perspective to think about what might be a new solution to an old problem.

Comment from audience:
For students it is useful not only to question cultural backgrounds, but also to get close to understanding what we know, what we can do and how we transfer things.

Berman:
We are always dealing with mediated information. Our perception is based on the amount of information that we remove.

Berger:
I am pessimistic. If you look at past dependencies and trajectories, then the amount of ideas you can use to come up with innovative ideas or models in planning at the university is extremely limited. There are particular reasons why specific things work in one place and usually they have to do a lot with history. Sure sometimes there are ideas that have not been thought of because of the burden of history, and coming from outside can be that kind of liberating experiences. But in many cases, it probably will not succeed.

Massey:
One thing is that the past is also an archive of possibilities that we do not see in the present. There is that potential of discovery in the past. And I really value that desire for empirical evidence. It can be a good provocation against speculation.

Berman:
It is necessary to be optimistic as an architect or urban designer because it is always about invention in relation to a future possibility.

Within certain fields the exchanging of information across territories occurs all the time, despite the fact that there might be very specific conditions. That kind of cultural exchange is absolutely critical for us to be able to move our disciplines forward.

Reicher to Paetzel:
There seems to be a lot of interest from international students in the transformation of the Emscher and to get knowledge that could be transferred to another region. What is your impression?

Paetzel:
The transformation is just engineering work. But the managing of the project is the biggest challenge.
We do not talk enough about power. The decision makers got the power and we got the knowledge but we have not got the capacity to offer them projects and give them the chance to create a better future.

II. From Walled City to Integrated City. International Summer School

Design Workshop 2018

TU Dortmund University, May 22th to May 27th

Worldwide, metropolitan regions are changing rapidly to absorb the urban growth expected over the next 20 to 30 years. While urban transformations in the global South will be dominated by irregular processes of urbanization, the formal regulatory frameworks operating in the global North need also to be questioned to ensure a more sustainable, just and vibrant urban future. Framed by the paradigm shift of the digital revolution, the anticipated impacts of climate uncertainty, and the aftermath of the post-industrial socio-economic regimes, time calls for innovative strategic responses. With emphasis on urban integration, the workshop thrives at addressing the complexity of these challenges through an integrated and transdisciplinary approach that allows for diverse, inclusive and multi-scalar design strategies on a specific context: Essen-Katernberg in the German Ruhr region.

With its 53 municipalities and over 5 million inhabitants, the Ruhr region is the largest metropolitan region in Germany. It went through diverse urban transformation processes, the most important of them was the structural change from a montane based economy to one based on education and services. The year 2018 marks a milestone in this process, as the last colliery in the area was shut down. On the one hand, several districts in the Ruhr region situated next to collieries have to deal with challenges opposed to this structural change, such as high unemployment rates, vacant plots, migration, shrinking. On the other hand, a need for visionary urban design adjustments, the demand for housing supply, innovative educational landscapes and attractive open spaces are standing for a chance to reshape the declining neighbourhoods into thriving quarters. The Katernberg district lies in the north-east part of Essen and next to the now famous "Zeche Zollverein", a coal refinery which closed in 1988. After an uncertain future, Zeche Zollverein was added to the UNESCO list of World Heritage in December 2001. Furthermore, a masterplan for its site was developed by Rem Koolhaas and the Office for Metropolitan Architecture OMA which was finished by 2010. The masterplan proposed a band around the site, which today is part of the regional-green-corridor network formulated in 2010 in the „Master Plan Emscher Landschaftspark". Moreover, this green band (or ring) constitutes a system of important green open spaces around the Zeche Zollverein ensemble and Essen-Katernberg. However, can a green connection by itself solve the problems for disadvantaged neighbourhoods?

In the "From Walled City to Integrated City" framework, a critical stance towards the masterplan is taken, and concrete questions on spatial design and strategic management of urban integration on a small-scale and its transferability pursue an update of the plan. For this purpose, four themes are to be developed with an overall focus on new alliances (or mixes) between them.

Education - How can a sustainable educational urban landscape be developed?
Designing educational landscapes
Public Space - What is the role of public space in urban integration?
Urban integration through public space design
Housing - Which future forms of housing can be developed?
Design concepts for innovative forms of housing in a spatially and socially integrated context
Work - How can local work be integrated on a small scale?
Integrating new forms of work (business, services, production), such as start-ups, into the urban fabric of a disadvantaged neighbourhood

The International Student Design Workshop 2018 was held at Zollverein Essen site and TU Dortmund University. It was intended to facilitate international cooperation among schools from Asia, Europe and North America. During the workshop, participants made plans to foster the development of interdisciplinary and multi-scalar design strategies. The sessions were supported through input from external consultants, and feedback from local stakeholders, communities and professionals.

The international workshop was carried out in cooperation with following universities:

TU Dortmund University, Faculty of Spatial Planning, Department of Urban Design and Land Use Planning
RWTH Aachen University, Faculty of Architecture & ILS Research Institute for Regional & Urban Development
University of Michigan, Taubman College
University of Virginia, Faculty of Architecture
Kobe Design University, City Planning Institute, Kansai
Kyoto University, Deptartment of Architecture and Architectural Engineering

Group 1 „The Triple F"

Ilenia Angarano, Adrian Bierholz, Kunheng Han, Genta Kanamaru, Weiwen Yang, Theresa Zorn

Designing educational landscapes

Conquering spaces by diverse food related activities favours an inclusive and collaborative community. Collaborative learning, sharing knowledge and increasing awareness are key aspects of urban farming as well as preparing food together that is connected to cultural programs and workshops. Vegetables, fruits and herbs can be cultivated in community and are later on prepared in the new activated spaces. Further activities as potential repair shops, green classrooms, cooking schools and workshops are forcing an inclusive way of learning. The Career Training Centre aims to activate a multifunctional centre in which spaces for learning are located. These spaces consist of a cooking school, a café/ restaurant, an open-plan kitchen and an outdoor space for chilling, meeting and eating.

Urban integration through public space design

The reprogrammed spaces, indoor as well as outdoor, create new neighbourhoods. By giving the opportunity to meet each other, common feeling is created and social cohesion is supported. Social meeting and communication points bring the community in Katernberg together.

Integrating new forms of work into a disadvantaged neighbourhood

In the context of the food topic several new small and middle-sized businesses develop that boost the local economy and enliven the heart of Katernberg. Already existing businesses are integrated to achieve synergy effects. The new network of diverse businesses, institutions and organizations foster the benefit from existing material and immaterial infrastructure.

The Field Testing Point can be presented in two ways: as the end of the culinary voyage or as the beginning of it. It includes the start-up centre Triple Z, where places for new businesses are located, and the Youth Centre, where young people can go and take part in several activities. Those two places, Triple F and Triple Z, collaborate and create multicultural events during which start-ups can present their new experimental products.

"Food"

To offer food to someone is an easy mode to establish communication. By discovering traditional and ethnic food, people create a connection that it is not only physical, but especially mental, because cooking is full of stories and stories are full of people.

Rediscovering the relationship with nature and its products. Source: Authors 2018

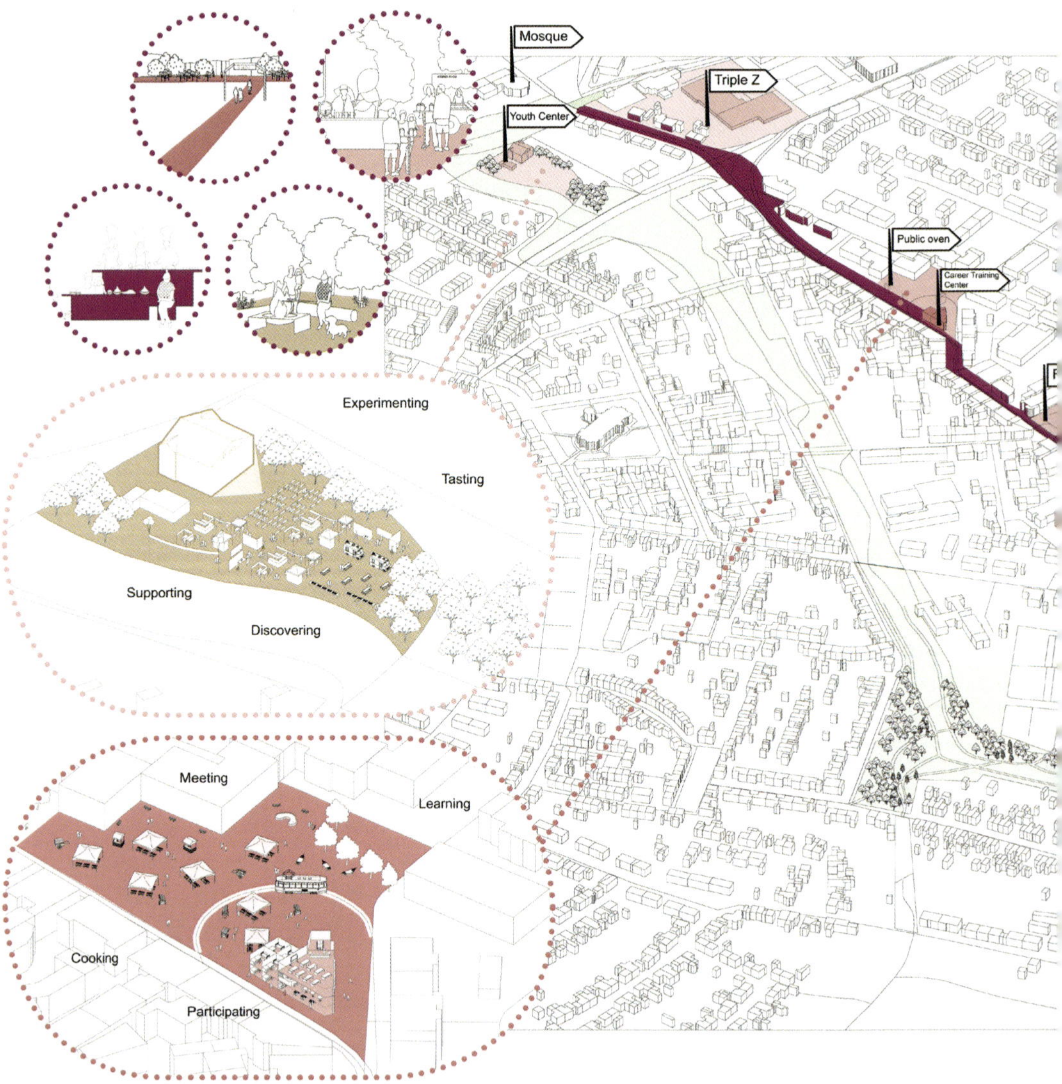

The entrance is presented as a place where people can rediscover the relationship with nature and its products like apples, pears and corn. The main street becomes a place of life. The market square has the assignment to magnetize the incomings. It presents itself as an open-air hall usable for events like social dinners, food festivals, and concerts. The Triple F and the Triple Z work together in the Field Testing Point. Thus, both sides can take advantage from each other: start-ups can have a feedback concerning their work and people participating in the multicultural events can have the possibility to experiment the "food of the future".

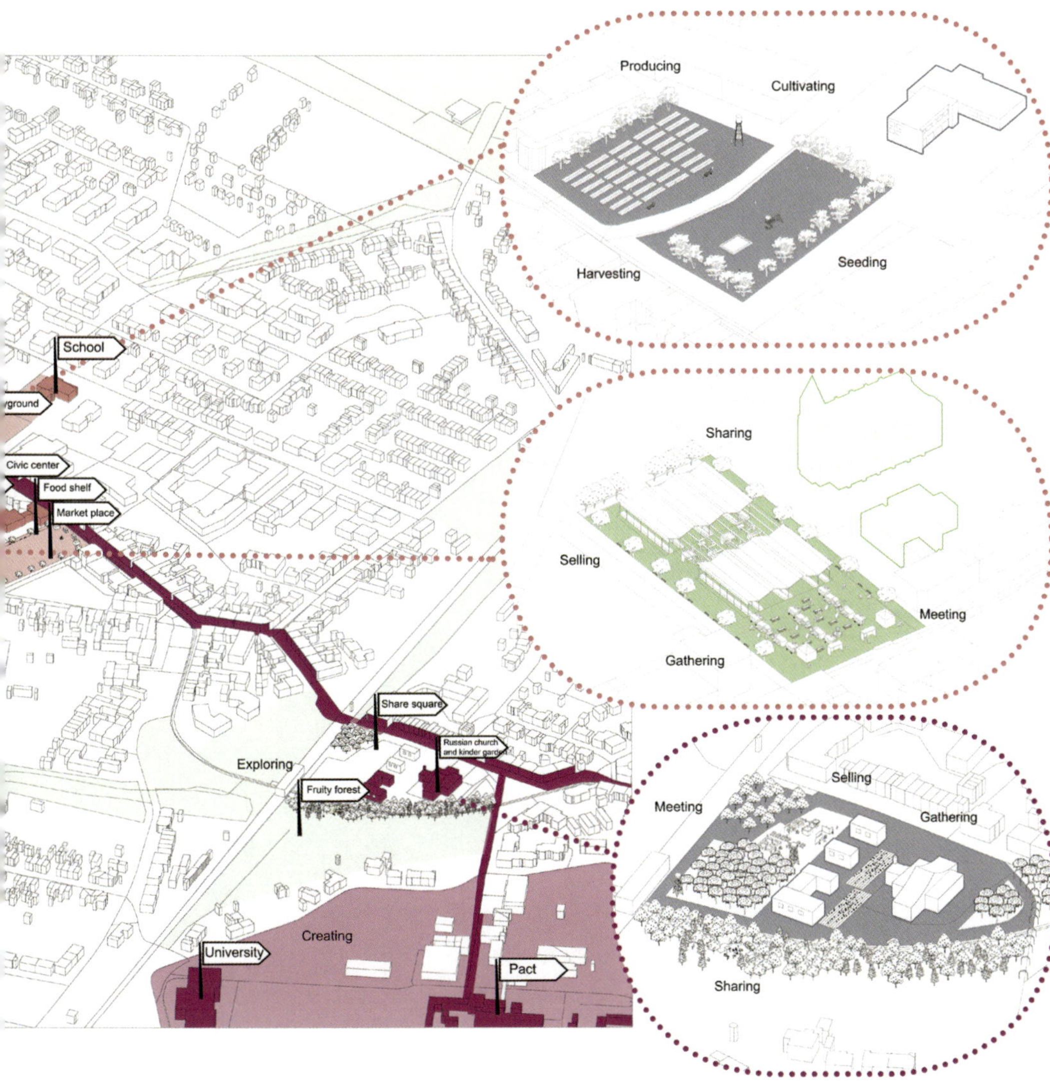

Group 2 „Nodes + Threads“

Joanna Bokuniewicz, Samuel Kirchner, Katarzyna Maliszewska, Shotaro Matsuyama

Nodes + Threads is a proposal for an enhanced network of existing and proposed pedestrian and bike trails that provides connections to local residents throughout the neighbourhoods adjacent to the Zollverein site, as well as to the greater region. It also introduces integrated programs through a series of accessible nodes that are directly connected to Zollverein and its surrounding neighbourhoods, which allows for a new kind of overlapping regional infrastructure. The project's two major components aim to cohesively break down physical and social barriers in the greater Essen area.

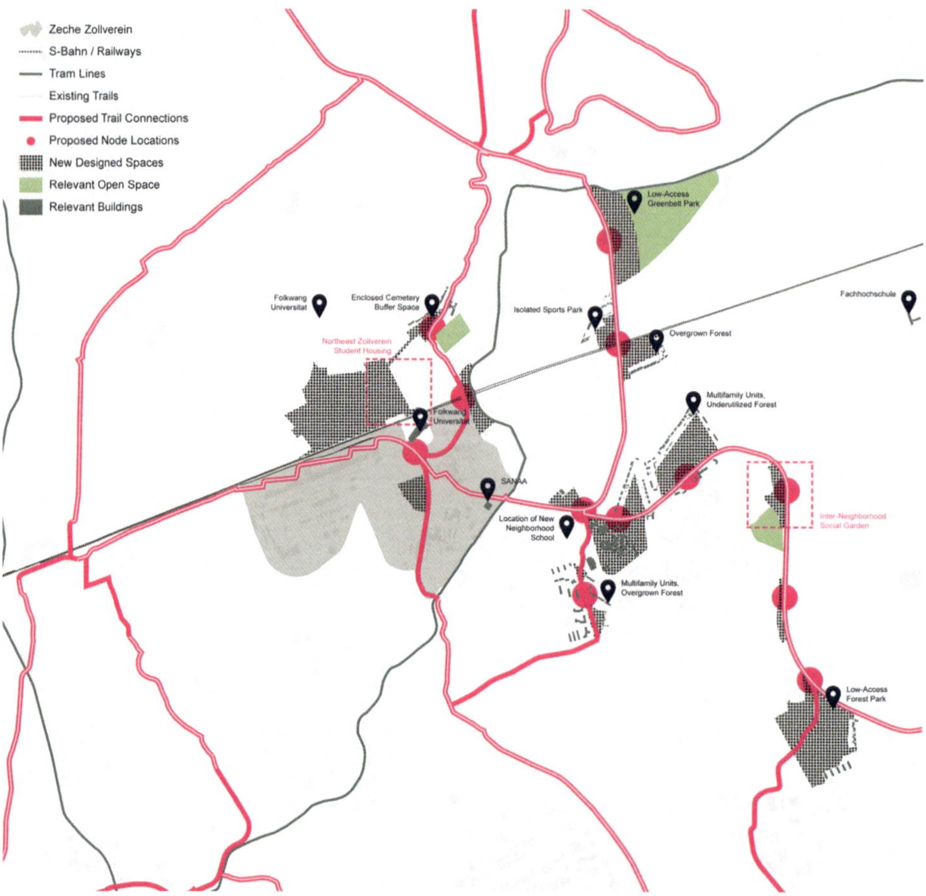

Enhanced network of existing and proposed pedestrian and bike trails. Source: Authors 2018

The project is based upon analysis of the relationships between the different neighbourhoods within the site, which – as the team quickly realized - were separated by an existing trail system. The team also perceived that this trail system had very infrequent access points which created a significant series of physical and social barriers. Further analysis showed that there were similar trail systems in many of the adjacent sites, but there was neither a legible connection between them nor any visible hints towards a relationship with a larger local or regional trail system.

In response to this seemingly fragmented system of trails, the team proposed a series of connections between the routes in the areas directly adjacent to Zeche Zollverein. The aim of this action was to form a legible and identifiable network of trails throughout the area, using Zollverein as a sort of "conduit" for connectivity and programmatic function. The team also saw the potential to connect this new system of routes not only with Essen, but with two larger biking networks in the Ruhr area, the Emscher Park Cycle Path and the Ruhr Area Circuit Road.

After connecting these disparate threads, the team identified important intersections or locations of overlaps between trails and other forms of transportation. Locations were deemed significant, for example, when a bicycle route crossed a suburban railway station, or two paths intersected near a school. These locations became the project's nodes, which were to be programmed according to the team's response to conditions in each site's immediate context.

The programmatic elements of the project's interventions were selected based upon the initial site analysis. The proposed connections between disconnected neighbourhoods aim to improve existing conditions not just through outright connectivity or accessibility, but through proposed additional functions that were selected according to the specific situation. Examples influencing programming include schools, universities, and residential neighbourhoods.

While analyzing the area's trails, the team identified three different typologies for existing conditions along trails, including 1) a raised condition, characterized by the presence of a mound 2 to 4 meters in height underneath the trail and dense vegetation on either side; 2) a lowered condition in which the trail was several meters below the adjacent context and surrounded by dense vegetation; and 3) a spanning condition, in which the trail was located on a bridge that spanned a street or stream below.

The programming of both nodes and threads also required a response to these three existing topographic site typologies. From them, the team identified three different appropriate programs that would correspond to each of the four project schemes, which included housing, education, business, and open space. Examples implied functions such as temporary housing, outdoor classrooms, pop-up stores and event spaces. The selection of these specific programs and their application toward existing site conditions led to the creation of a set of components that could be located at nodes either together or individually, depending upon the specific situation. All of the components are located along existing and proposed bike lines, and often include additional facilities for cyclists and pedestrians, such as bike stands, ramps, or bike sharing stations.

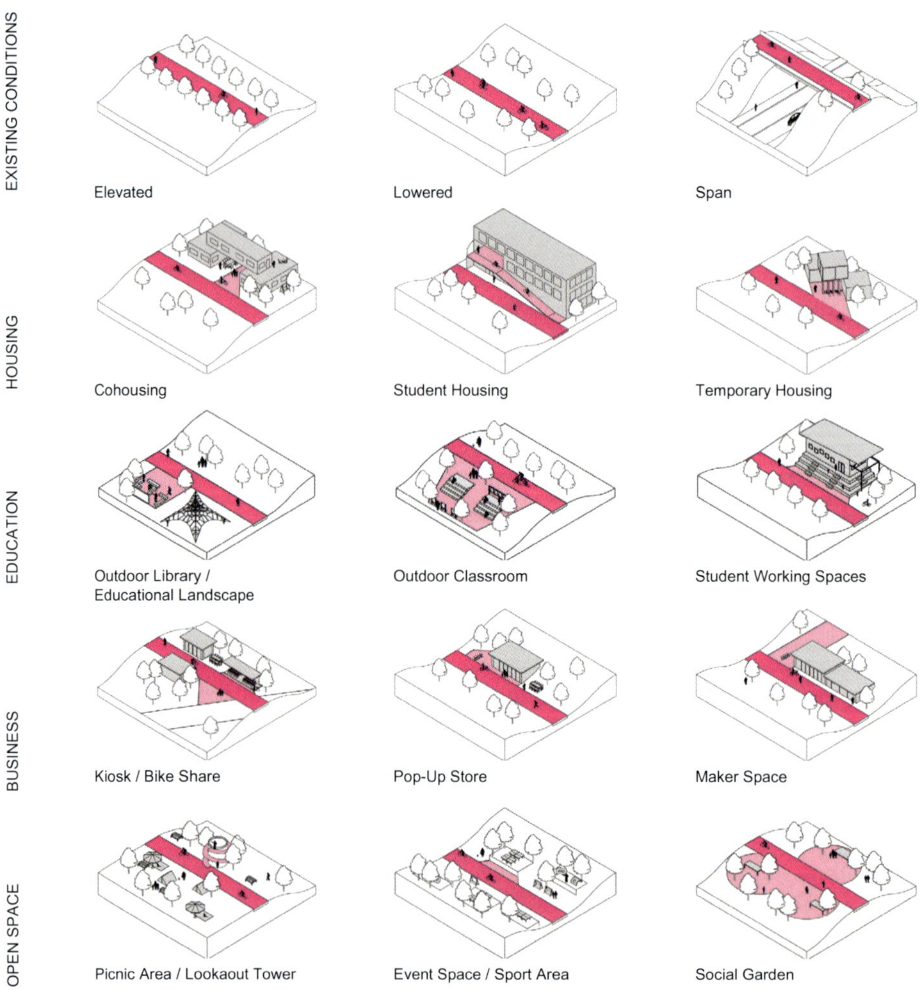

Design components of the Nodes + Threads network. Source: Authors 2018

Next, the team envisioned two different scenarios for new programs along the new network, utilizing the aforementioned components and responding the idiosyncrasies of the selected sites. In the first scenario, the functions of student housing, sports, event spaces, and pop-up shops are arranged in a site just north of the Zollverein. The site was selected for these uses due to its proximity to the university, as well as the SANAA building. In the second scenario, a social garden, bike share structures, temporary housing, and business kiosks are located in a space that is currently a dense but otherwise overgrown forest space adjacent to the bike path.

Nodes + Threads attempts to create new proposals for different functions that can be placed, mixed, and adjusted according to the needs of the immediate surrounding context, as well as to the needs of the larger city and region. By placing various functions along cycling paths, they become programmatically activated and connected, allowing them to provide more than just the traditional function of linear movement. The potential for overlap and integration of these functions, as well as their access to and relationship with a larger network not defined by traditional automobile infrastructure, leads to speculation of a new kind of linear urbanism.

Group 3 „ART-HUNT“

Marei Asche, Katsuya Komatsu, Mathias Leuthe, Kelsey Ryan, Maximiliane Sattler

The analysis of the site showed that meeting points play an important role in the daily lives of the residents in the area around Zeche Zollverein but that there are not provided enough places to meet people from the neighbourhoods. From the analysis performed by all the teams, we understand that the neighbourhoods surrounding Zollverein have unique characteristics and that they are composed of a diverse population but at the same time lack connections. Taking this disconnection of the existing neighbourhoods and their physical centre, Zeche Zollverein, as a starting point for an overall idea, the main focus of concepts and strategies should be the aspect of connection.

Showing a way to reflect and uncover the characteristics and potentials of each neighbourhood leads to the concept of using "art" as the connecting link between both the individual neighbourhoods and Zollverein and its surrounding districts. Existing projects, such as PACT and the sculpture forest at Zeche Zollverein, provide the opportunity to be a bridge between the separated areas. Furthermore, the concept makes use of existing potentials of "art" in the neighbourhoods as well by redefining art as a learning process, a connector, and an identifying feature. By expanding the scope of what is considered art, art can take many forms, from the high art you can see in museums expressed through visual mediums such as paintings, sculptures or photography, to daily activities that are reflective of local culture and the character of a district. Therefore, art is capable of existing at many scales, temporal states, and through various actors; the existing artistic activities can be rendered visible through the concept of ART-HUNT. Art lies in the eye of the beholder, has become the foundation of the envisioned artistic network, ART-HUNT creates a network of existing and emerging art that provides flexible, interactive and a wide range of uses. Here art takes form in teaching how to cook a meal, growing a public garden, the production of handcrafts or spontaneous street performances. Art is integrated into many parts of the daily routines in Essen's surroundings and possesses the ability to unite people of different backgrounds through shared appreciation and participation. Additionally, by redefining art as described before, the controversial and discussed term "art" is used to accentuate the existing distinctive features of the neighbourhoods and to attract tourists in a slightly provocative way.

The concept of ART-HUNT is based on a structure of goals and strategies. All the goals are assigned to the overall objective of connecting the residents of the surroundings with each other and to link Zeche Zollverein with the surrounding neighbourhoods.

The first goal is to identify and strengthen the characteristics and identification points of each neighbourhood. Open participation is one step to include the residents into the approach and to meet their needs. Providing a platform for local and regional artists through dynamic and interactive processes is another step to include the residents and their conception of art. ART-HUNT aims to be inclusive to all who wish to participate, and the freedom that the new definition provides enables a diverse range of creative people to be part of the curation process. The development of an information sharing system such as a volunteer board of residents and communication devices such as websites and mobile applications creates a designated platform for the continual curation of ART-HUNT.

Network of learning, sharing and creation. Source: Authors 2018

For the second goal, attracting the visitors and tourists to come into the neighbourhoods, the existing art has to be uncovered and appropriate places for new art have to be found and used. The locations of art can be distributed throughout the neighbourhoods in spaces that are potentially abandoned buildings, open spaces and/or vacant lots. The place should provide a distinct identity and further amenities in addition to the art. In this sense the art spaces range from permanent locations like gardens, cafés, educational playgrounds or sharing libraries, while others can be more ephemeral, existing for maybe a day or a week. Performances in community parks, galleries in vacant buildings or offering a cultural experience through a community dinner can be part of these art locations.

And finally, it is necessary to provide tools and a network for tourists and residents to find the art-places and to take part in the actual ART-HUNT. This is reached by creating the ART-HUNT app, which provides different ways to participate in the hunt. With maps, localized bike-rents and bus stations, potential overnight accommodations and an event calendar it is possible to explore the surroundings on one's own and in a flexible way.

The third goal is to attract the residents to use the Zollverein area which is currently not seen as a meeting point for the surrounding residents. Therefore, interesting spots for the residents should be created at Zollverein. Daily amenities and places that invite people to linger should be placed thoughtfully in the area of Zollverein. On top of that a strategy to bring people from the surrounding areas to the inside of Zeche Zollverein is to open up underused places within Zollverein and its local conditions as local learning experiences for the local residents and schools. By creating designated spaces that provide educational, leisure, and recreational opportunities residents can be attracted into the space that is currently seen as walled off to the neighbourhoods.

The last goal is to create a more or less tangible connection of the neighbourhoods and of the Zollverein with its surroundings. By using ART-HUNT logo on the participating art-places and the signs leading to the individual locations identifiable paths are created, and ART-HUNT is not only accessible by using the app but can also be joined spontaneously. By an identifying logo, a connection can be experienced by the residents as well as by the tourists. Another connecting tool is presented in form of the ART-HUNT app and its interactive features. The included chat option facilitates to connect the residents among each other, to get to know new people and to arrange to meet and stay in touch. At the same time the app plays a connecting role by navigating through the neighbourhoods and connecting the different art-pieces and locations.

ART-HUNT not only highlights the creative activity happening in the neighbourhoods surrounding Zollverein, but it bolsters the existing economic environment. By drawing curious visitors and tourists out into the community, this prompts the opportunity for more people to visit local restaurants, shops, hotels, and use other services. ART-HUNT enables money that was previously concentrated inside the "walls" of Zeche Zollverein to be more evenly distributed in the neighbourhoods, therefore, increase the economic stability and overall quality of life of the neighbourhoods.

Highlighting the creative activity happening in the neighbourhoods surrounding Zollverein.
Source: Authors 2018

Group 4 „Blue Threads“

Rebekka Maria Graw, Tomomi Honjo, Niklas Philipp Käding, Tsvetelin Tsvetanov, Junxi Wu

Analysis

First the analysis refers to the east-west parts of Zollverein and the surrounding quarter called Stoppenberg. In the area there are two schools and eight kindergartens, which are necessary for the supply of educational institutions. The educational institutions are spread over the whole area. There are no other educational institutions such as extracurricular education. The area is equipped with the basic educational institutions but there are no additional ones like for example scientific institutes. For the further supply structures are located in three business places. One is a little shopping mall in a neighbourhood, which is close to Zollverein. In the east, there is an industrial estate, which includes for example a machine construction company. Additionally, the Gelsenkirchener Strasse is a shopping street with flower shops, bars, bakery etc. The buildings are organized in a block structure. Otherwise, in the whole area there are especially family houses, duplex houses and a row structure. All neighbourhoods are embedded in green space like parks, playing grounds or allotment gardens. In the south green space, there is located a canal named "Stoppenberger Bach" and a pump station of Zollverein, which is important for the concept Blue Thread.

Concept

It is known that the historic coal mining industry is well connected with water, which needed to be managed for the safety of underground activities and was also used for cooling the coal seams. However, from the experience of our field trip, we could observe that currently, water in this area has been highly controlled and isolated from people, with canals, water pipes and pumps; in many parks they are fenced off. We further found out that due to the former mining activity water needs to be constantly pumped out from the underground to prevent the area from being flooded, which represents the idea that water has long been treated as a threat or risk to the region, responded by the high intention to separate urbanity from it. From a hydrological perspective, the area is connected by this unique water situation, with scattered pumps protecting not only communities, but also the Zollverein site.

Unfortunately, the historical connection between water and this area, and the

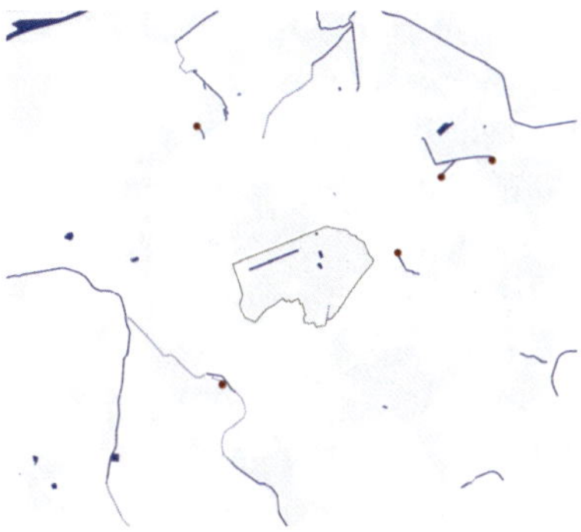

Existing Situation
Pumps and scattered water detached from Zollverein

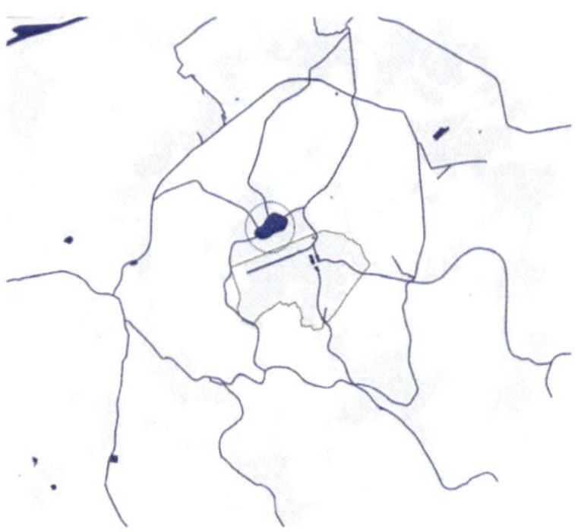

Proposed Network
Connecting existing water and leading to Zollverein

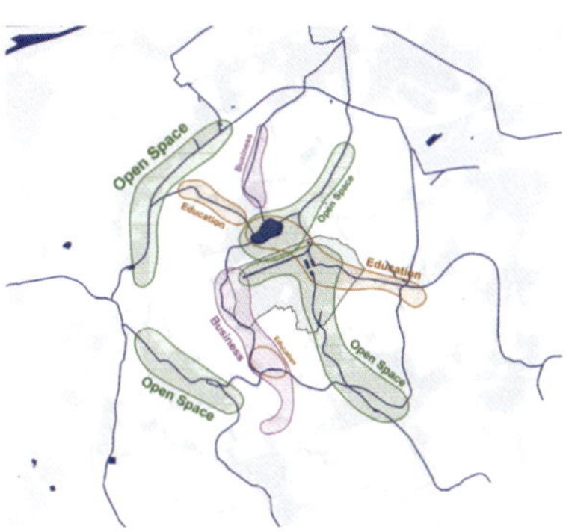

Vision
Water as improvement strategy along corridors

Envisioning water as improvement strategy. Source: Authors 2018

hydrological connection between Zollverein and the neighbourhood are not recognized or acknowledged by the public. One of the examples is the water experience in Zollverein, with its iconic pool and cookery acting as one of the attraction to tourists, but not accessible by the communities. In response to this situation, we think that it is crucial to put forward the advocacy that water is an important part of the heritage assets and should play an important role in integrating this area. Also, water should be repurposed to serve for the benefit of the communities and should be accessible and celebrated by the public. Following this provocation, we propose a water network connecting existing water in the parks, taking advantage of the low-lying area in the green ring around Zollverein and the stable resource from pumped-out underground water. Also, by extending into the patchwork of open space around Zollverein, it can become a diverse and connected water system, encompassing different types of canals, wetlands and water infrastructure. The network can act as a spatial guidance of paths connecting Zollverein and the communities and improve quality of life along the corridors. We envision that this water network can influence the area in many ways by providing better environment for residents and local business, promoting knowledge about water management, acting as a potential mitigation to the risk of flooding, and tying the area as an integrated whole.

In the present concept, two types of water systems have been developed. One type has a linear and the other one a flat structure. A common feature of both is that they are not rigid entities. The illustrations show two different scenarios for each location. The first scenario shows the respective location in a state in which the water pumps work as usual. The second scenario takes into account a possible condition in which the pumps do not pump out the groundwater. It not only illuminates the positive side of water, but also includes a negative risk of water landscapes in the conceptual planning. The respective locations will continue to be usable in the event of the occurrence of both scenarios. The resulting flexibility makes the locations particularly attractive. One example is the canal in the community. The water strengthens the attractiveness as a residential location. In the normal state, the water serves as a design element here. People can sit relaxed on the shore or go for a walk and enjoy nature in the middle of the neighbourhood. In a flooded state, the space around the canal offers a whole new use. People can use it as a place to swim, get together with other people and learn about the laws of nature and especially water. This place, therefore, serves as an integrating place. The images of the canal in the commercial street and the canal on the disused railway line show no flooded scenarios, as flooding almost excludes any use. As another important location we present a solution for the Zollverein area in the face of a Water Plaza at the end of the Education Path. Reshaping the existing ponds into pools, small plates on different levels are created, thus orienting the visitors into the landscape. A water pump is masked behind a picturesque artificial waterfall, serving the role

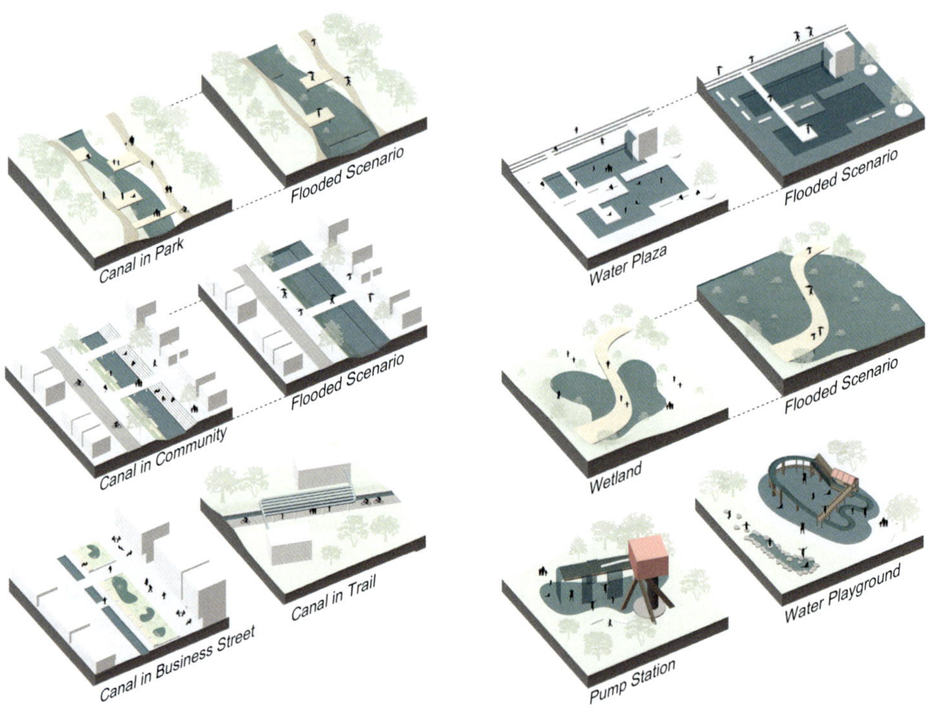

Thinking flooded scenarios within the design components. Source: Authors 2018

of one of the main attractions. The whole ensemble is framed with stairs and functions as a big emergency pool in case of flooding. The urban furniture and the elevated bridge create alternative surfaces for walking so that the plaza is still accessible to the public. With this scenario in mind, more areas for gathering the water are needed. As a solution, possible wetlands are localized, creating not only emergency pools, but places of high climate benefits as well. The light bridges spreading through them serve both as beautiful promenades and transit routes for emergency access. In the points of the pump stations water canapés can create an interactive zone with possible swimming pools and baths, which are not present for the community at the moment. The pumped water has an estimated temperature of 30 degrees Celsius, so it can function all year round and create steam baths in the winter season. Furthermore, water playgrounds are spread around the educational areas and in proximity to the kindergartens in order to raise awareness of the importance of water and its impact on environment from young age. The sketched perspectives illuminate and reinforce the idea of the different situations.

In conclusion, we envision that this water network can influence the area positively in many ways by providing a better environment for the residents and local business, promoting knowledge about water management, acting as a potential mitigation to the risk of flooding, and tying the area as an integrated whole.

Group 5 „Common Ground“

Franziska Berlin, Nakamura Fumiya, Lennart Jording, Annika Rüther, Niklas Trudrung, Jinshnai Zhang

What is a Walled City?

Zollverein Park is an isolated patch in the city and a development enclave which treats international creative intelligences. It is owned and operated by an entity which functions in broader scale and with big intention.

Zollverein Park was rich in culture which was deeply embedded in the vicinal neighbourhood and used to be the centre and hub of the society in the industrial era. Along with the wane of industrial glory, the peripheral region was rebuilt in the context of industry legacy. Different ethnicities and groups inhabit different neighbourhoods; the city is divided by the regional transportation infrastructure. Communities are fragmented and some of them are lacking of the developmental motivations and opportunities.

The masterplan proposed by OMA aims to build a clearly defined peripheral redevelopment zone, which functions as a preservative wall for industrial legacy.

Concept: Common learning ground

Learning from the precious tradition of polycentric urban patterns, we know that the collaboration between multi-municipalities played a key role to forge the Ruhr miracle. With appreciation of local culture and history, we advocate a more

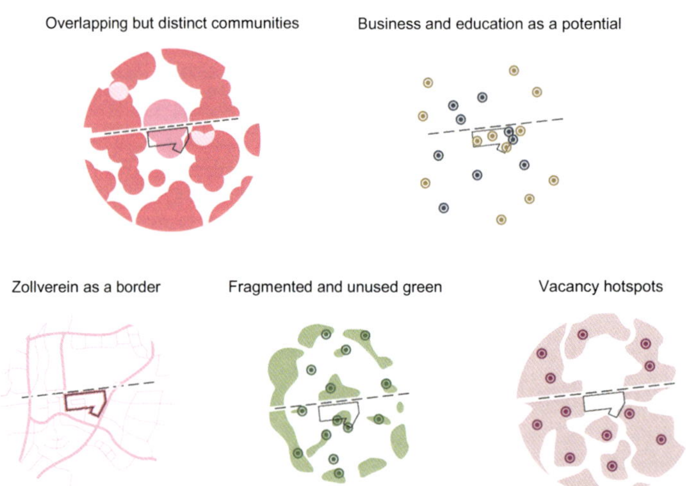

Facilitating communication, mutual learning and shared study. Source: Authors 2018

Three steps towards a common learning ground.
Source: Authors 2018

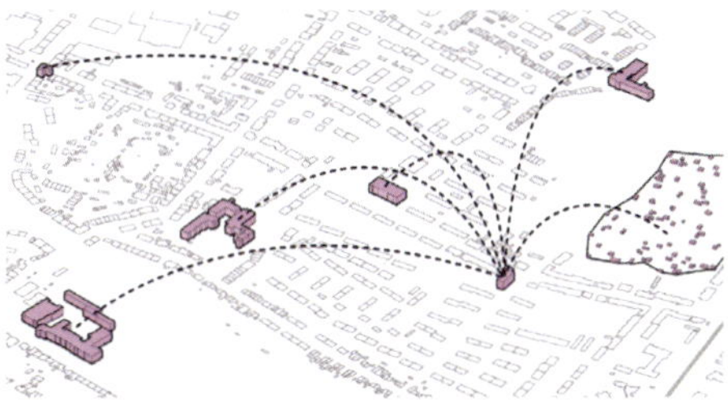

Ground level activation

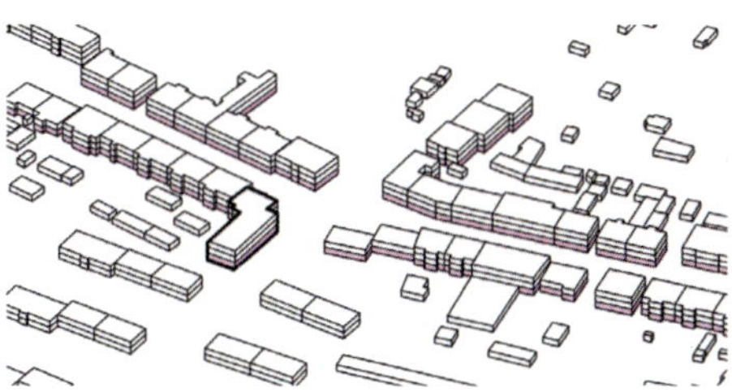

Stakeholder involvement

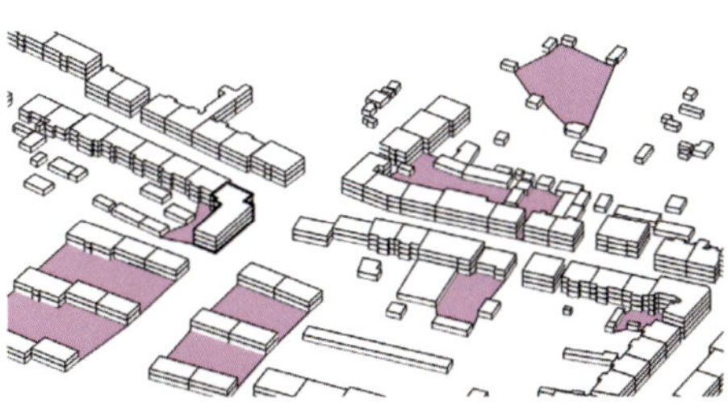

Backyard programming

integrated society through collaboration between Zollverein Park and nearby communities; we promote the concept of common learning ground to facilitate communication, mutual learning and shared study.

How could social integration succeed? In the vision of common learning, we think that it should work in multi-levels and multi-scales, from personal meeting to community conversation. The Common Ground can be defined by private study and public meeting, by all kinds of civil facilities or semi-public space.

As an international innovative hub, Zollverein Park is playing a great role in building a globally learning society. The regional green belt acts as regional potential linkage space, attracting nearby neighbourhoods to participate and share common knowledges. As to community level, we identify main public buildings, educational institutions and other potentially underutilized space. This space will be the potential space for hosting community learning ground.

We also consider including different age groups to forge life-long learning. In terms of education, through the participation of different schools and colleges and universities, kids and undergraduates can learn from each other via joining the same games, social interactions and conversations.

In order to spatialize the common learning ground, a learning network landscape will be established on the supports of various stakeholders. On the one hand, we advocate strengthening the international influence by inputting local learning image and people. Through the extension of the University of the Arts, the installation of a preschool at Zollverein Park and a reprogramed SANAA building, we try to image mix-used scenarios in which an inclusive society can be achieved.

On the other hand, we advocate the limited measurements of renovating existing community. We image the social collaborations could be organized by certain QNGO's (Quasi Non Governmental Organizations) not through PPP mode (Privat Public Partnership).

We think there are natural demands for connection existing in the society, and the space we provide aims to stimulate the potential moments of encounter, power the conversation between people with different skills or interests.

Conclusion

The concept of Common Ground aims to coordinate the uneven geographical distribution of social resources, stimulating daily encounter and mind exchange through a highly recognizable urban learning landscape. The hubs will accommodate organizations and organizers who are dedicated to match social supply and demand in terms of building a common sense of learning society.

Daily encounter and mind exchange through a highly recognizable urban landscape.
Source: Authors 2018

Group 6 „Fragmentation Integrated“

Hares Ahrari, Adham Amar, Marie Kötterheinrich, Ryuhei Maruoka, Keigo Nakamura, Jabarin Nur

The analysis was carried out on six different themes. The first was on the housing and density typology in which we noticed the different housing structures we had, like large-scale multi-family houses, single-family houses, houses in a row and block structures, which usually offer living space for different social groups.

The second was on the different forms of businesses, scales and their emissions. The prevailing use of the land plots of our area is the residential use with the exceptions of some small-scale businesses scattered all over the area. There is also a large car logistic centre that occupies a large area of the site. Interesting for us was noticing the impact of the different forms of businesses on the liveability of the neighbourhoods. We noticed that the neighbourhoods that have some kind of small-scale businesses feel more liveable and safer than the neighbourhoods next to the car logistic centre.

The third part of the analysis was about education. The site has different forms of educational facilities like primary schools, a nursery and a youth support centre which is run by the church.

The fourth part was on the public and the semi-public spaces. Due to the many boundaries and borders of the site, there are many not walkable areas. The semi-public spaces especially in the social housing projects are ignored and not taken care of, like for example the spaces between residential buildings in an area where a big part of the residents are elderly people; those spaces are completely ignored and the residents do not have any physical access to them. For the publicly accessible spaces, the situation did not differ much from the previously mentioned ignored spaces as a large sum of the actually well-kept spaces like those to be found at Zollverein have little to no benefit to the surrounding community despite the fact that the site has a significant touristic value. As for the large forest: it acts only as a physical barrier separating Zollverein from the surrounding neighbourhood and is only used as an area for hiking and leisure activities.

The fifth topic was the different boundaries and borders we found at the site. There are not only physical borders dividing the site spatially but also social borders between the communities. The physical borders of the site consist of two railway tracks, the physical wall of the car logistic centre and the big forest.

The sixth topic was about the social situation of the site. The community is diverse with so many migrants and people from different cultural backgrounds; also, a big part of the community is inhabited by elder generations. Furthermore, members of the community were interviewed and asked about the social structure, the community relationships and the problems they found in the area.

Concept

Based on our analysis results we developed our goals and vision for the area. In order to efficiently develop goals for the area, we concentrated our conclusion to the four main regions forming the site, which are the Zollverein site, the forest, the semi-public spaces and the car logistics centre.

Our Goals

(1) ZOLLVEREIN: benefit from the surrounding community and actually use the large open areas and some of the unused buildings

(2) FOREST: open up the forest and utilize some areas of the forest to serve the community

(3) THE SEMI-PUBLIC SPACES: give the community a sense of belonging to those spaces by creating rooms inside those spaces to increase the communication between community members

(4) THE CAR LOGISTICS CENTRE: integrate the centre in the area instead of treating it as only a physical barrier that needs to be overlooked

Our methods of achieving those goals

(1) ROOM: a small room with flexible use that could be given to the public to conduct different activities in. Those small rooms could either be rented with or without cost. The individuals have the choice to decide on how many rooms to rent as well as on their use

(2) a partial or a full reuse of the area

(3) create a landmark or a highlight

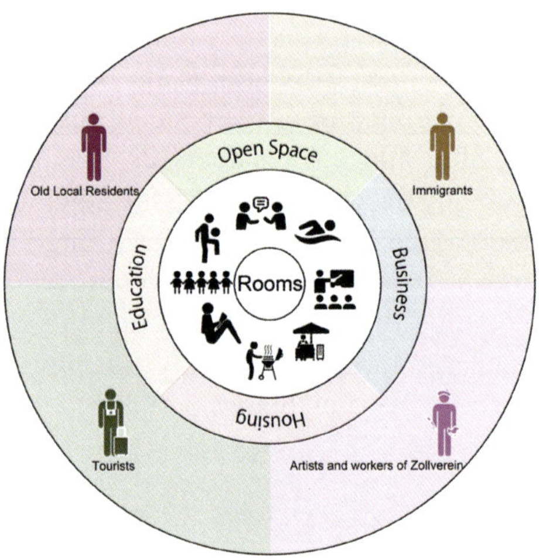

Building blocks of the ROOMs
Source: Authors 2018

Implementation at Zollverein

As mentioned before our goal for the area was to initiate that Zollverein would give something back to the surrounding neighbourhoods. From our analysis, we noticed the problems in those neighbourhoods, which were a high unemployment rate especially in the migrant community and a lower chance for the youth to join high schools, which would lead to a lower chance of finding jobs later on.

At the Zollverein site, we used two of the above-mentioned methods.

(1) In the large open green lawns and car parking areas, our ROOM method is used in which the public rent those rooms to do whatever they want to do with them. As mentioned before the rooms are of a flexible use but there are also some rooms with for example predetermined uses that could be changed from week to week or month to month to host different forms of exhibitions and fairs.

(2) The unused buildings, floors or even some spaces in the buildings will be used as spaces for education and community support, which help people finding jobs and start their careers.

Implementation in the forest

Our goal for the forest was to open it up and not treat it as a barrier but as an opportunity. For the forest, we used the ROOM and the highlight methods. We placed highlights spread out through the forest; those highlights could be a public theatre or a big open-air market. Around those highlights, we placed clusters of the mentioned small flexible use units (ROOMs) which the people could rent for different activities.

The forest as opportunity for flexible uses. Source: Authors 2018

Implementation in the semi-public space

Our goal was to create a sense of belonging and to decrease the social gap between the people; that is why we used the ROOM method. We implemented this method in two spaces. First, in the social housing projects most of the residents were elderly people. In this model, the "ROOMs" were to be placed at the beginning of the semi-public space and would be used as areas where people could sell things, exchange used stuff, handmade products and food. Second, in the social housing projects which were mostly inhabited by migrants and had a big social gap. Our approach to these spaces was similar but instead of placing the units at the beginning of the space, they were placed in the centre to encourage and spark communication between the residents. Those “ROOMs” would not fully occupy the semi-public space but only took a part of it; the rest were intended to be conventionally used, like for leisure, relaxing and as play grounds for kids. The actors running this model in the semi-public spaces would be elected members of the community and the residents. They should be responsible for handing out the areas and organizing the whole renting processes. It is also important to mention that these “ROOMs” were to be given out to the residents who want to rent them for free; if there was no demand by residents, then they could be rented from externals who would rent them at low price; the rental fee will then be collected by the elected residents who will distribute it later among the residents on a monthly or an annual basis.

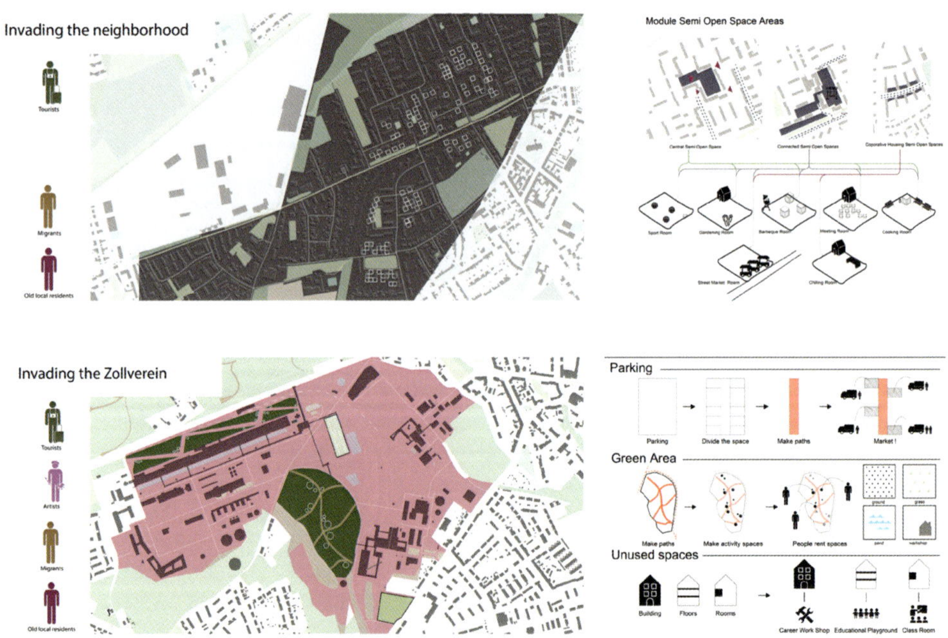

Sense of belonging through ROOMs in the neighbourhood. Source: Authors 2018

The car logistic centre occupies a large space of our area hence adding more to the division of the site. Our goal was to open it up to the public. We proposed to build a platform on the large chimney that is found in the centre to be used as a highlight and an attraction where people can take a look over the city and do different activities on top of it.

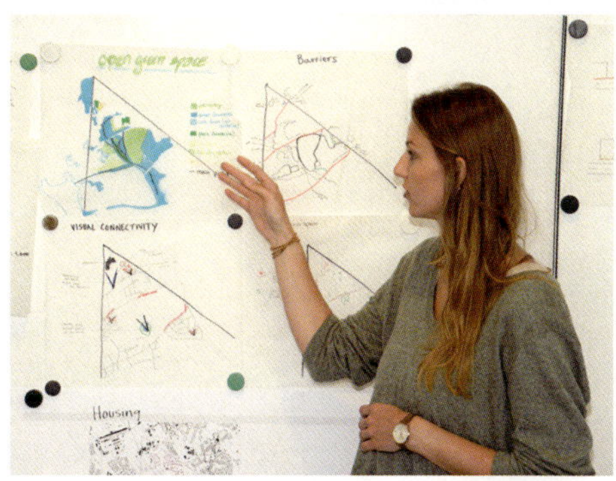
open green space
Barriers
VISUAL CONNECTIVITY
Housing

III. Outlook

Dimensions of Urban Design and Integration

Christa Reicher

Since the coal and steel industry started in the Ruhr region more than 150 years ago, the region changed from an agricultural landscape to a highly densified region where the economic development of the time had an important influence on the spatial structure of the region. It's called "Agglomeration Ruhr" because the borders between its 53 cities are not relevant in terms of settlement space. Since the 1960s, diverse population, diversified economy, mixed use, landscape as a backbone, and innovative infrastructure were in the centre of the transformation process of the previously industrial Ruhr region. The challenges of urban development and urban design are linked to different urban dimensions and they have to address one important topic: integration. Urban development and urban design should focus on spatial aspects; however, they are also challenged to address social change to offer new spatial, creative and strategic solutions for the city and the region of tomorrow.

In this context, the urban neighbourhood or quarter as a framework or a scope for actions seems to be one of the most effective scales, at least in this region marked by several administrative borders. These borders where based on political and administrative decisions without any relationship to the spatial reality. By dealing on the quarter level, tailor-made concepts can be designed and implemented, while following an integrative approach and involving citizens and the local stakeholders. Nevertheless, many of these challenges still need concepts or strategies on the municipal and regional scale.

Regarding the question on how to deal with the challenge of integration, five theses can be elaborated which are at the centre of this contribution:

(1) Spatial conditions have an important impact on social integration

Quarters are to be designed and shaped in a way that they offer motivation and assistance in counteracting segregation tendencies and downward spirals in already disadvantaged neighbourhoods. Urban settings are permanently changing conditioned by societal shifts, and migration accelerates this transformation. The Ruhr region has experienced this for about 150 years. In the phase of industrialization, workforce from different countries in Europe were recruited by actors in the Ruhr region; they were needed, they were necessary, to secure the growing wealth. Since the era of the montane industry in the Ruhr region differentness became something like the "usual". Although the Ruhr region in its present form could only emerge through immigration and the "differentness" almost naturally belonged to this region, concentration processes that lead to disadvantaged neighbourhoods became visible.

From the perspective of urban design and urban and regional development, the spatial laws and frame conditions and their ability to influence integration are of particular importance. The possibility of participation in society is an important prerequisite for integration; it can be promoted or obstructed by the spatial conditions. Urban planning must, therefore, deal with this desire for demarcation and at the same time aid for integration. There is a link between an urban environment and the promoting effect of urbanity. Not least because of this, cities can prevail as "integration machines", provided they meet the requirement of "places for arrival" (see Saunders 2011). An urban context – if understood as spatial and social density combined with mixed-use – seems more suited to naturally pick up the "other" and the "different" and connect it with the familiar. The other is at the same time a call for change and thus a chance for innovative architecture, for different forms of spatial organization and new concepts for urban and neighbourhood development.

(2) Public space plays a key role within an integrative urban development

Debates around the future of cities and urban growth are inseparable from the debates around public space and public sphere – both in the theoretical discourse and in the practice of city planning. Architects and city planners, politicians and civic activists, researchers, and public relations specialists are all aware of the significance of public space as backbone and image of a quarter, a city, or a whole region.

The grammar of the city as well as the rules of public space are changing permanently according to the needs of the society and the demographic situation. The change of public space is often related to a loss of experience. Spaces and parks are not used and designed as in former times; new types of public space were developed without any rule. At first sight, the public space seems to be a space with a lot of problems and deficiencies: loss, gentrification, privatization, commercialization, neglect and vandalism. But there are also quarters with high-end designed and staged space. The range of public space is the result of different social tendencies, of a pluralization of lifestyles as well as of growing and shrinking quarters side by side.

The public space within cities and regions has become to be the main concern and focus of architects and planners when thinking of integrated development. In this framework, public space can get expressed through various forms and faces.

It could be communicated as urban green, as temporal installation of space and program as well as long-term valuing of streets and spaces.
It does not have to be an either-or policy, but a better balance between public interests and private enterprises will have to be worked out in order to ensure that the invaluable free joys of public life will be available citywide to all of us in the cities of the present and of the future.

(3) The places for social infrastructure can function as a crucial catalyst of integration

Speaking about social infrastructure, schools and kindergartens come to mind which are not always open to the neighbourhood and not at all linked to the community. Creating a kind of network with these different institutions of education and opening them to the public would be a great challenge.

Variety and diversity became socio-political keywords in the present.

Children spend significantly more time in educational institutions today than they did thirty years ago. Contemporary educational architecture does not only change in its internal organization and design, but also in its relationship to the outside world. Various educational and cultural institutions are connecting to learning and educational landscapes within a neighbourhood, the city itself is increasingly becoming a place for learning and the transitions between the educational site and the adjacent urban space are considered with greater care because educational institutions should generally have a greater effect on their environment. Whether schools, educational centres or higher education institutions - in the context of urban redevelopment and urban renewal, they increasingly serve as engines for future neighbourhood development and increase the attractiveness of the entire city.

When I speak of social infrastructure this includes - in addition to the educational and meeting places - the sports and leisure facilities and their associated infrastructure – they are considered as an integration factor no. 1. The restoration, renewal or expansion of sports facilities, football fields or other sports and leisure facilities play a central role in this context.

(4) Technical innovations and digitalization may have a significant impact on urban space and integration

Intelligent urban systems and social networks improve the efficiency of resources and support social participation. One of the main questions is how digitalization influences the physical space.

Digitalization does improve the access to knowledge and information. Digital networks support the identification of people with the surroundings and political participation.

Neighbourhood concepts must, therefore, incorporate the technological possibilities as an option while at the same time keeping an eye on their relevance for the physical space.

(5) Beside formal plans, there is a need for informal planning strategies and integrated planning approaches

In times of ever more complex transformation processes of cities and neighbourhoods, concepts, strategies and instruments that go beyond planning and organizing and that take informal approaches into account are needed.

In the course of this development new planning methods become necessary which take complexity, non-linearity, uncertainties and self-organization into account. Whereas traditional urban planning aimed at establishing a final stage of development and preferably precise spatial design for a future urban quarter, the central task of an informal urban design is the active shaping of the development itself. The discussion about the formal urban form is not in the foreground, but rather the question about possibilities of city activation and urban use.

It is well known that city planning can only be successful by adequately integrating social, ecological and economic concerns. Integrated urban development concepts connect in a balanced way "bottom-up" approaches with "top-down" approaches, in a strategic development management.

In this sense, planning is getting ever more complex and calls for a multi-scale perspective of urban design. The future of the city is the link between the region and the quarter, and urban concepts of the future need to be developed with an interdisciplinary approach with a strong focus on a multi-levelled social and spatial integration.

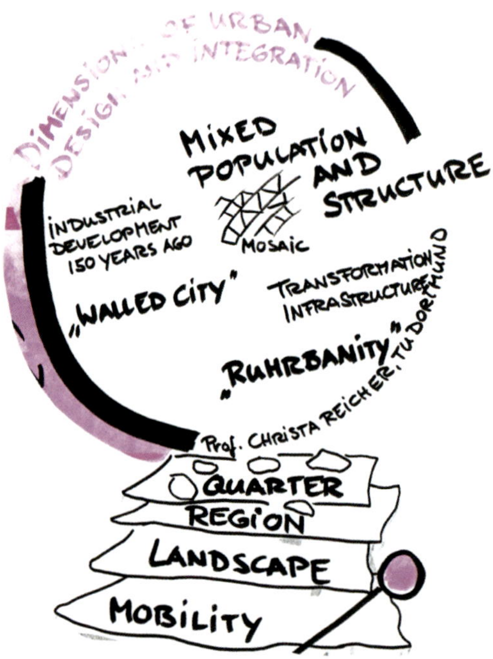

© Antje Rieder, Graphic Recording

About the Authors

Assoc. Prof. María Arquero de Alarcón, University of Michigan

María Arquero de Alarcón is an Associate Professor of Architecture and Urban and Regional Planning and Director of the Master of Urban Design at the University of Michigan Taubman College. She is a founding principal at MAde Studio.

Dr. Asad Asadzadeh, University of Bonn

Asad Asadzadeh studied urban and regional Planning at the University of Teheran, Iran, and holds a PhD (-Ing) in Urban Planning & Land Management from University of Bonn, Germany. He is currently working as a postdoc at the Department of Urban Planning and Land Management, University of Bonn.

Fabio Bayro-Kaiser, RWTH Aachen University (TCR-Team)

Fabio Bayro-Kaiser, Research Associate and Doctoral Candidate at the Chair of Urban Design and the Institute for Urban Design and European Urbanism of the Faculty of Architecture at RWTH Aachen University. Prior to this, he worked as a research associate at the Department of Urban Design and Land Use Planning in the Faculty of Spatial Planning at TU Dortmund University, and as an architect at "m.Kaiser: architecture and planning" in Tarija, Bolivia.

Prof. Dr. Stefan Berger, Ruhr-Universität Bochum

Stefan Berger is Professor for Social History and Director of the Institute for Social Movements at Ruhr Universität Bochum. He is also Executive Chair of the Foundation History of the Ruhr and Honorary Professor at Cardiff University. Before, he was Professor for German and Comparative European History at the University of Manchester, and Professor of Contemporary History at the University of Glamorgan.

Assoc. Prof. Scott Campbell PhD, University of Michigan

Scott Campbell is Associate Professor of Urban Planning and Director of Doctoral Studies in the Taubman College of Architecture and Urban Planning at the University of Michigan. He studied urban planning (PhD, MCP) at the University of California, Berkeley, and Environmental Earth Sciences and German at Stanford (BAS), plus additional education at the Technische Universität Berlin (through DAAD) and the Freie Universität Berlin.

Marielly Casanova, University of Duisburg-Essen

Marielly Casanova teaches at the University of Duisburg-Essen where she finished her doctoral thesis in 2017. Casanova's previous work in research and teaching extends to Columbia University, NY and the ETH Zurich. In addition, Casanova worked as a project manager and head of the department of the Public Domain Projects - Municipality of Chacao in Caracas, Venezuela.

Robin Chang, TU Dortmund University

M.Sc., Research Associate, PhD Candidate at the Department of European Planning Cultures, Faculty of Spatial Planning, TU Dortmund University. Studied spatial planning at TU Dortmund University and environmental planning at the University of Northern British Columbia.

Dr. Assoc. Prof. Mona El Khafif, University of Virginia

Mona El Khafif is an Associate Professor at UVA School of Architecture and Principal of SCALESHIFT, received her professional degree in architecture from the RWTH Aachen and her doctoral degree in urban design from the TU Vienna. El Khafif taught at the TU Vienna at the Institute for Urban Design, Tulane University, California College of the Arts and the University of Waterloo, where she co-directed the school's DATAlab.

Dr. Carmen Maria Enss, University of Bamberg

Carmen M. Enss studied architecture and heritage conservation in Weimar, Munich, Trondheim and Bamberg. In 2013, she received a PhD in architectural history. Her main research interests as a research assistant with the Centre for Heritage Conservation Studies and Technologies (KDWT) at Bamberg University are history of city planning and architectural conservation.

Maarten Gheysen, KU Leuven

Maarten Gheysen (B) is architect, urban designer, and PhD-candidate at the Department of Architecture of the University of Leuven. Since 2006 he has been teaching in the design studios at the Faculty of Architecture. He started his professional career in the office of Xaveer Degeyter Architects and afterwards started working as urban designer in a municipal development agency (Leiedal).

Stephanie Haury, BBSR

Stephanie Haury studied architecture and urbanism at the University of Karlsruhe and Barcelona. Planning practice in architectural offices in Amsterdam (NL), teaching and research practice in urban design at the University of Siegen. Since 2009 researcher at the Federal Institute for Research on Building, Urban Affairs and Spatial Development (BBSR) in Bonn.

Päivi KataikkoGrigoleit, TU Dortmund University (TCR-Team)

Päivi KataikkoGrigoleit is an architect, assistant professor at the Department of Urban Design and Land Use Planning at the Faculty of Spatial Planning at TU Dortmund University and a partner of REFLEX architects_urbanists in Essen, having previously worked as an architect in several offices in Finland, Sweden, Germany and Holland.

Prof. Dr. Kiyoko Kanki, Kyoto University

Kiyoko Kanki is a professor at Kyoto University, Japan. In 1992 and 1993 she was visiting researcher at the Chair of Landscape Ecology and Planning, RWTH Aachen. Former chairperson of Rural Cultural Landscape Sub-Committee in Architectural Institute of Japan. Since 2015 she has supported the branch office of Kyoto City to promote the upgrade of illegal development areas in the Fushimi Green Belt.

Silja Kessler, TU Dortmund University

M.Sc. Silja Kessler studied spatial planning at TU Dortmund University, Faculty of Spatial Planning, graduation 2018. In 2015 Erasmus Exchange at Blekinge Tekniska Högskolan, Sweden. Since 2017 she has been working as a student intern at s.t.e.r.n NRW, Neighborhood Management, and she is NGO member in 'Die Urbanisten', Dortmund.

Dr. Andreas Kipar, LAND, Landscape Architecture Nature Development

Andreas Kipar studied landscape architecture at Essen University GHS from 1980-84, and architecture and urban development at the Milan Polytechnic from 1989-1994, where he has taught Public Space Design since 2009. His landscape architecture office LAND, which has offices in Italy, Switzerland, and Germany, offers regional, national, and international clients services in designing livable spaces for the world of tomorrow.

Prof. Dr.-Ing. Theo Kötter

Theo Kötter has been Professor of Urban Development and Land Readjustment at the University of Bonn since 2003. He worked for several years for urban development companies and therefore has extensive experience in urban redevelopment and development as well as in infrastructure planning. In current research projects he is concerned with new approaches for a resilient and climate-adapted city, for a socially equitable and sustainable land policy, for prospering urban regions, as well as for the transformation of settlements and infrastructure in times of demographic and economic change. He also deals with the interactions between urban development and real estate markets.

Prof. Dr. Hisako Koura, Kobe Design University

Hisako Koura is Professor of Kobe Design University, and a certificated urban planner. She holds a PhD from Osaka University in 1996, and from 1997-2015 was the Associate Professor of Graduate School of Engineering, Osaka University. Her academic work on the urban landscape was awarded the AIJ prize of 2012. She is committee member of World Heritage in the Agency for Cultural Affairs of the Japanese Government.

Dr. Christian Lamker, TU Dortmund University

Christian Lamker studied planning and environmental law in Dortmund and Auckland. He graduated in spatial planning in Dortmund in 2010 and worked as research assistant, lecturer, school manager study & tuition and study coordinator for TU Dortmund University, between 2010 and 2016. In between, he went as Visiting Research Fellow to RMIT University, Centre for Urban Research, Melbourne and finished his PhD in spatial planning in Dortmund. Currently, he serves as research assistant and lecturer for spatial planning and planning theory.

Prof. Jonathan Massey, Dean, University of Michigan

Architect and historian Jonathan Massey is Dean and Professor at the Taubman College of Architecture and Urban Planning at the University of Michigan. Previously Dean of Architecture at California College of Arts, he was also the Laura J. and L. Douglas Meredith Professor for Teaching Excellence at Syracuse University. Massey is a co-founder of the Aggregate Architectural History Collaborative.

Prof. Dr. Angela Million, TU Berlin University

Angela Million, née Uttke, urban designer and urban planner. She is Professor of Urban Design and Director of the Institute of City and Regional Planning at TU Berlin. Before she was a researcher at the Institute of Urban Affairs (Difu) in Berlin. She studied urban design and planning in Germany, Spain and the US. She holds a PhD from TU Dortmund.

Sarah Müller, TU Dortmund University (TCR-Team)

M.Sc. Sarah Müller is a spatial planner and research associate at the Department of Urban Design and Land Use Planning of the Faculty of Spatial Planning, TU Dortmund University. She completed her master‘s degree in 2018 with a focus on urban design.

Prof. Martin Murray, University of Michigan

Martin Murray is Professor of Urban Planning in the Taubman College of Architecture and Urban Planning at the University of Michigan. In Taming the Disorderly City (Cornell University Press) and City of Extremes (Duke University Press), he critically engages with the spatial landscape of Johannesburg after apartheid.

Jun. Prof. Dr. Jan Polívka, RWTH Aachen University, ILS Dortmund (TCR-Team)

Jan Polívka is Junior Associate Professor for Sustainable Housing Development at the Faculty of Architecture, RWTH Aachen University, and Deputy Head of the

Built Environment Department at the Research Institute for Regional and Urban Development (ILS) in Aachen, Germany. He is an associate at rha assoziierte Architekten Stadtplaner Ingenieure in Dortmund, Germany.

Dipl. Designer Antje Rieder, Freelancer

Antje Rieder works as a graphic recorder, does live visualizations and illustrations. She graduated from the University Wuppertal and holds a degree in Industrial Design. She works for companies, advertising agencies, publishers and TV stations. In the year 2000, she designed the Music Award „Eins Live Krone". As a graphic recorder, her main interest is to support people by her visual work. In meetings or conferences, she makes real time drawings and takes visual notes, thus providing a reflection surface of a special character to the participants. Thus, all vignettes in this publication were developed live at the TCR conference.

Prof. Christa Reicher, RWTH Aachen University (TCR-Team)

Prof. Christa Reicher is Head of the Chair of Urban Design and Director of the Institute for Urban Design and European Urbanism of the Faculty of Architecture at RWTH Aachen University. She is founder of the architecture office rha reicher haase architects in Aachen and in Dortmund. Prior to this, she was Professor for of Urban Design and Land Use Planning at TU Dortmund University and for Urban Design at the University of Applied Science, Bochum, and worked as a research associate at RWTH Aachen.

Assoc. Professor Kris Scheerlinck PhD, University of Leuven, Belgium

Architect and Urban Designer, Associated Professor at Department of Architecture. He ran his own research and design practice with projects in Barcelona, Milan, Madrid, New York, Antwerp, Ghent and Brussels. He currently directs an international research project called 'Streetscape Territories' and promotes related PhD research projects. His expertise is on the analysis of public-private gradients in urban projects and their related models of accessibility, permeability and proximity, analyzing territorial boundary delimitation in real life projects in Europe, South and North America and Africa. He is the Head of the 'Urban Projects, Collective Spaces & Local Identities' research group.

Prof. Jörg Schröder, Leibniz University Hannover

Architect and urban planner, full professor and Chair for Regional Building and Urban Planning of Leibniz University Hannover LUH; Director of the Institute of Urban Design and Planning IES. Graduated from Technical University of Munich TUM; from 2001–12 he used to teach and research with TUM.

Dr. Sabrina Sposito, Leibniz University Hannover

Sabrina Sposito is Territorial Planner, PhD. Academic education in University of Naples Federico II, Italy, II level Master's Diploma in Planning and Sustainable Design of Port Areas, PhD in Urban Design and Planning. Currently she is guest scientist at the Institute of Urban Design and Planning, Faculty of Architecture and Landscape Sciences of the Leibniz University Hannover. Previously: Postdoc Research Fellow at the Institute for Advanced Studies on Science, Technology and Society of Graz; Visiting PhD candidate at the University of Southern California of Los Angeles.

Assoc. Prof. Geoffrey Thün, University of Michigan

Geoffrey Thün is Associate Professor of Architecture and Associate Dean for Research and Creative Practice at Taubman College University of Michigan. He is a founding partner at RVTR. His awards include R&D Awards, Architizer A+ Award, ACSA Faculty Design Award, Royal Architectural Institute of Canada Award of Excellence in Architecture, Professional Prix de Rome in Architecture, Young Architects Forum Award from the NY Architecture League, and a Canadian Architect Award of Excellence.

Erik van Daele PhD, University of Leuven, Belgium

Erik van Daele is lecturer and studio master at the Department of Architecture. He received a master in architecture at the current KUL Faculty of Architecture Campus Ghent, a master in theory and history and a master in urbanism and spatial planning at the KUL, extension studies program at the AA school in London. From 1998 till 2014 he was associate in the design firm uapS, Paris. In his PhD (hybrids as open signifiers) he explored small scale design challenges in the dispersed city. His design focus is on disturbances, imperfection, uncertainty, weakness as a quality, appropriation and designing possibilities.

Assoc. Prof. Kathy Velikov, University of Michigan

Kathy Velikov is an architect, Associate Professor at University of Michigan's Taubman College, and President of ACADIA. She is founding partner of the research-based practice RVTR, a recipient of the Architectural League's Young Architects Award and the Canadian Professional Prix de Rome in Architecture.

em. Prof. Dr. Michael Wegener

Until 2003, Michael Wegener was Director of the Institute of Spatial Planning and Professor at the Faculty of Spatial Planning of TU Dortmund University. In 1988 and 1989 he was Professor at the Department of Civil Engineering of the University of Tokyo. Since 2003 he is a partner at Spiekermann & Wegener (S&W) Urban and Regional Research, Dortmund.

Dr. Dominik Weiß, University of Bonn

Dominik Weiß studied spatial planning at TU Dortmund University and economics at the University of Karlsruhe. PhD in economics, Martin Luther Universität, Halle-Wittenberg. He is currently working as a postdoc researcher at the Department of Urban Planning and Land Management, University of Bonn. Former positions: guest lecturer at the Vietnamese-German University in Ho Chi Minh City (Vietnam), economist at the Department of Urban Economics at the Halle Institute of Economic Research (IWH) and Project Manager at Wüstenrot Haus- und Städtebau GmbH, Ludwigsburg.

Assoc. Prof. Claudia Wigger, University of Michigan

Claudia Wigger is a lecturer in architecture at University of Michigan Taubman College. She studied architecture at TU Berlin and has 16 years of experience as a licensed architect, through collaborations and consulting, such as Foster and Partners, Sauerbruch Hutton in Berlin, RAD in Hong Kong and Hamilton Anderson Associates in Detroit.

Assoc. Prof. Claire Zimmermann, University of Michigan

Claire Zimmerman is Director of Doctoral Studies in Architecture at the Taubman College of Architecture and Urban Planning at the University of Michigan. Her current book project is Building the World Capitalist System: Albert Kahn and the Architecture of Industrial Detroit.